"Finally! Too often mentoring gets fl̲ ̲. But here Dr. Guyotte and Dr. Wolgemuth have curated and edited a volume that offers conditions for possibility, opportunities for mentoring to matter differently, and possibilities for liberation. Most appreciated is the ways the collection of chapters, images, poems, and mentoring moments reveal mentoring from the perspectives of students, junior and senior scholars all committed to *Philosophical Mentoring in Qualitative Research: Collaborating and Inquiring Together*."

— **M. Francyne Huckaby,** *Associate Dean,*
TCU School of Interdisciplinary Studies,
Professor of Curriculum Studies,
TCU College of Education and Center for Public Education, USA

"This edited collection begins with the premise that mentoring is a relational practice and that such relations need not be hierarchical in order. This necessarily inventive orientation shifts the terrain upon which mentoring is understood, even practiced: power relations disperse; subject distinctions (of mentors, mentees) blur; newly creative practices emerge. It is here that this important text takes seriously the notion of mentoring as a philosophical practice, one that necessarily entails a way of relational living that exceeds the traditional bounds of the academy and the ordering of method. As a result, this text will be useful – even inspiring – to mentors, mentees, and mentor-mentees alike."

— **Aaron M. Kuntz,** *Frost Professor of Education and*
Human Development, Department Chair, Counseling,
Recreation, School Psychology, Florida International University, USA

"*Philosophical Mentoring in Qualitative Research: Collaborating and Inquiring Together* addresses the diverse forms that qualitative mentoring might take, inviting readers to consider mentoring as philosophical practice. In this edited collection, authors describe the productive possibilities of engaging with uncertainty, discomfort, and the unexpected within relational approaches to mentorship that value caring and ethical connections with humans, non-humans and the more-than-human. Thought-provoking and refreshing, the book engages readers in envisioning new possibilities with respect to how qualitative researchers are always becoming – whether novice or experienced, or in formal or informal spaces. Filled with poems, narratives, conversations, and images, this collage of chapters and mentoring moments provides a dazzling array of alternative approaches to thinking about and engaging in mentoring relationships. Read on and savor!"

– **Kathryn Roulston**, *Professor of Qualitative Research,*
Department of Lifelong Education, Administration,
and Policy, University of Georgia, USA

Philosophical Mentoring in Qualitative Research

With contributions from advanced, early career, and emerging qualitative scholars, *Philosophical Mentoring in Qualitative Research* illuminates how qualitative research mentoring practices, relationships, and possibilities of inquiry and teaching come to life under different mentoring philosophies.

What we can know in and about the world is inseparable from our approach(es) to knowing with and in it. And how we mentor in qualitative research matters to what we can know and do as qualitative inquirers. Yet, despite its importance, mentoring is rarely conceptualized as a practice inspiring or inspired by philosophy. This edited book opens a needed space for thinking about mentoring as a philosophical practice. Its thoughtful chapters and artful "mentoring moments" draw on critical, feminist, new materialist, post-structuralist, and other philosophies to make visible, interrupt, reflect, deepen, and expand mentoring practices within the qualitative community revealing what we can know, do, and become through them.

Philosophical Mentoring in Qualitative Research sensitizes readers to mentoring as a philosophical practice. As such, it is essential reading for students and researchers in qualitative research and higher education interested in mentoring practice and humanistic research values.

Kelly W. Guyotte is Associate Professor of Qualitative Research at the University of Alabama, USA, and Coordinator of the Educational Research Program. Inspired by her background in the visual arts, her research interests include gender and equity in higher education, artful inquiry practices, STEAM education, and teaching/mentoring in qualitative inquiry.

Jennifer R. Wolgemuth is Associate Professor of Educational Research at the University of South Florida, USA, and Director of the Graduate Certificate in Qualitative Research. Drawing on critical theories, she explores inquiry as an agential process that investigates and creates the lives and communities to and for which researchers are responsible.

Philosophical Mentoring in Qualitative Research

Collaborating and Inquiring Together

Edited by
Kelly W. Guyotte and
Jennifer R. Wolgemuth

Routledge
Taylor & Francis Group

LONDON AND NEW YORK

Cover image: Choose Life, by Lisa Love Whittington, 2021

First published 2022
by Routledge
4 Park Square, Milton Park, Abingdon, Oxon OX14 4RN

and by Routledge
605 Third Avenue, New York, NY 10158

Routledge is an imprint of the Taylor & Francis Group, an informa business

British Library Cataloguing-in-Publication Data
A catalogue record for this book is available from the British Library

Library of Congress Cataloging-in-Publication Data
Names: Guyotte, Kelly W., editor. | Wolgemuth, Jennifer R., editor.
Title: Philosophical mentoring in qualitative research : collaborating
   and inquiring together / edited by Kelly W. Guyotte and
   Jennifer R. Wolgemuth.
Description: Abingdon, Oxon ; New York, NY : Routledge, 2022. |
   Includes bibliographical references and index. | Summary: "This
   book looks at the ways that mentoring relationships can can become
   a practice of philosophy. By looking at mentoring experiences and
   relationships through a new materialist framework, the chapters
   highlight the intrinsically relational nature of mentoring. Through that
   relationality, mentoring practices can disrupt and push against hierarchies,
   boundaries, normativity, authority and power. This is increasingly
   important as mentoring roles come to prominence in a neoliberal
   academic environment. Suitable for students and researchers in
   qualitative research and higher education interested in mentoring
   practice and humanistic research values"—Provided by publisher.
Identifiers: LCCN 2021036484 (print) | LCCN 2021036485 (ebook) |
   ISBN 9780367900885 (hardback) | ISBN 9780367900892
   (paperback) | ISBN 9781003022558 (ebook)
Subjects: LCSH: Qualitative research. | Mentoring. |
   Mentoring in education.
Classification: LCC H62 .P463 2022 (print) | LCC H62 (ebook) |
   DDC 001.4/2—dc23
LC record available at https://lccn.loc.gov/2021036484
LC ebook record available at https://lccn.loc.gov/2021036485

ISBN: 978-0-367-90088-5 (hbk)
ISBN: 978-0-367-90089-2 (pbk)
ISBN: 978-1-003-02255-8 (ebk)

DOI: 10.4324/9781003022558

Typeset in Adobe Caslon Pro
by Apex CoVantage, LLC

CONTENTS

ACKNOWLEDGMENTS

We are indebted to all the people, relationships, experiences, and exchanges that made this book possible. Mentoring is a relational practice and relationships endure, continuing to leave their marks, well beyond in/formal meetings and departures. The possibility of this book, our coming together, our becoming scholars (of qualitative inquiry), were all enabled by our mentors, be they people, experiences, texts, and more. While we could never name all the people we have treasured as mentors on our own academic journeys, we acknowledge those who stand out to us now, in the context of curating this special edition, as invaluable and influential. These are people who both supported our becoming scholars and showed us what good, relational mentoring makes possible (alphabetically): Drs. Sarah Bridges-Rhodes, R. Brian Cobb, Tracie Costantino, Bob Dedrick, Kathleen deMarrais, Ellyn Dickmann, John Ferron, Melissa Freeman, Cliff Harbour, Mirka Koro, Aaron Kuntz, Tess Lea, Jessica Nina Lester, Barbara Pamphillon, Kathy Roulston, Steve Thoma, and Jessica VanCleave.

Of course, this book would not be possible without the authors and artists who contributed chapters and shared their mentoring moments. We are grateful for your honest, reflective, and brilliant work. We appreciate your engagement with each other as reviewers and your generous and generative feedback. We also thank Hannah Shakespeare at Routledge for her enthusiasm, guidance, and support throughout the book proposal and writing processes. Finally, we are beyond grateful to Dr. Lisa Love Whittington, who generously allowed us to feature her artwork on the cover.

We owe a special debt of gratitude to the two fabulous graduate assistants at the University of Alabama, Carlson H. Coogler and Shelly Melchior, whose collective attention to detail during the proofing, copyediting, and indexing processes well exceeded our hopes and expectations. We thank them both for the considerable time and effort they gave to readying each chapter for submission.

Finally, as current and former chairs of the American Educational Research Association's Qualitative Research Special Interest Group (AERA QR SIG) Mentoring Committee, we want to thank all committee members, participants, volunteer mentors, and SIG executives who supported and participated in the SIG's many mentoring initiatives. The QR SIG provided the space to both enact and reflect on mentoring and was fertile ground for philosophizing our mentoring practices. We thank everyone, past and present, who gives generously of their time and energy to mentoring (as a philosophical practice).

CHAPTER CONTRIBUTORS

Diane Austin, University of South Florida

Diane Austin is Instructor in the School of Information, College of Arts and Sciences, at the University of South Florida. She is a PhD student in the College of Education with teaching and research interests in youth services in libraries and school media, distance learning, digital story-telling, library outreach, makerspaces, digital literacy, and instructional technology.

Scottie Basham, University of South Florida

Scottie Basham is a PhD student in educational leadership and policy studies at the University of South Florida. She is interested in disrupting normative structures impacting voice, silence, and all that lies in between that often create expected, replicated results especially within educational settings.

Kakali Bhattacharya, University of Florida

Dr. Kakali Bhattacharya is a multiple award-winning professor at the University of Florida housed in the Research, Evaluation, and Measurement Program. She is the 2018 winner of AERA's *Mid-Career Scholar of Color*

Award and the 2018 winner of AERA's *Mentoring Award* from Division G: Social Context of Education. Her co-authored text with Kent Gillen, *Power, Race, and Higher Education: A Cross-Cultural Parallel Narrative*, won a 2017 Outstanding Publication Award from AERA (SIG 168) and a 2018 Outstanding Book Award from the International Congress of Qualitative Research. She is recognized by *Diverse* magazine as one of the top 25 women in higher education.

Nicole Bowers, Arizona State University

Nicole Bowers is a PhD candidate in the Mary Lou Fulton Teachers College at Arizona State University. Her work interlaces philosophy and methodology to explore socioecological issues with a focus on educating for uncertain futures. Her interests have led her to promiscuously engage across disciplines. Her most current research explores approaches to inquiry of past epistemology in the context of environmental and sustainability education in the Anthropocene.

Susan O. Cannon, Mercer University

Susan O. Cannon is Assistant Professor of Early Childhood/Middle Grades Education at Mercer University. Her scholarship traverses the fields of mathematics and statistics education, qualitative inquiry and teacher education. She reads and thinks with feminist, critical, new materialist, and posthumanist theories in order to work the boundaries of concepts and fields. Her work has been published in *Qualitative Inquiry, International Review of Qualitative Research, Taboo: The Journal of Culture and Education,* and *Education Policy Analysis Archives*.

Jonathan M. Coker, University of South Florida

Jonathan M. Coker is a PhD candidate in curriculum and instruction at the University of South Florida, where he also earned a Graduate Certificate in Qualitative Research. He is interested in teacher working conditions, LGBTQ+ issues, as well as critical and post-structural approaches to inquiry.

Dionne Davis, University of South Florida

Dionne Davis is a PhD candidate in educational leadership and policy studies at the University of South Florida, Tampa Campus. Her research interests include leadership in two-way immersion schools, equity audits, and retaining Black women as school leaders.

Maureen A. Flint, University of Georgia

Maureen A. Flint is Assistant Professor in Qualitative Research at the University of Georgia where she teaches courses on qualitative research design and theory. Her scholarship braids together her interests in the theory, practice, and pedagogy of qualitative methodologies, artful approaches to inquiry, and questions of social (in)justice, ethics, and equity in higher education spaces. Maureen's work has been published in *Qualitative Inquiry, Cultural Studies ↔ Critical Methodologies, The Review of Higher Education*, and the *Journal of College Student Development*.

Ryan Evely Gildersleeve, University of Denver

Ryan Evely Gildersleeve is Professor of Higher Education and Associate Dean for the Morgridge College of Education at the University of Denver. His transdisciplinary research involves cultural analyses of educational opportunity, discursive analyses of education policy, and ontological analyses of higher education's purpose, role, and relevancy in contemporary conditions. Gildersleeve earned his MA in higher education and organizational change and his PhD in education from UCLA. He is a graduate of Occidental College.

Anna Gonzalez, University of South Florida

Anna Gonzalez is a doctoral candidate at the University of South Florida, where her interests span a wide range of topics pertaining to qualitative methodologies, philosophies of inquiry, and teaching and learning.

Kelsey H. Guy, University of Alabama

Kelsey H. Guy is a doctoral student in the Educational Research program with a focus in qualitative research at the University of Alabama. She also works as a full-time instructor of Italian in the Department of Modern Languages and Classics. Her research interests include non-traditional transcription methods, qualitative research pedagogy, and the second-language acquisition and teaching of less commonly taught languages.

Samantha Haraf, University of South Florida

Samantha Haraf is a doctoral candidate studying teacher education at the University of South Florida. Her research agenda revolves around equity and social justice issues in education, and teacher learning in relation to these issues. She also studies and works in the areas of teacher leadership, student voice, and clinically based teacher education.

April M. Jones, University of Alabama

April M. Jones is a doctoral student at the University of Alabama, pursuing an educational research PhD in qualitative methodology. April is also a licensed clinical social worker with over 15 years of experience in the non-profit sector. Her research interests and paradigms are influenced by her identity as a woman of color in America and her experiences as a social worker interacting with professionals and youth/families in the juvenile justice system. Qualitative research affords April the opportunity to mean-ingfully explore the experiences of marginalized, underrepresented, and his-torically oppressed populations.

Mirka Koro, Arizona State University

Mirka Koro (PhD, University of Helsinki) is Professor of Qualitative Research and Director of Doctoral Programs at the Mary Lou Fulton Teachers College, Arizona State University. Her scholarship operates in the intersection of methodology, philosophy, and sociocultural critique. She is the author of *Reconceptualizing Qualitative Research: Methodologies without*

Methodology (2016), co-editor of *Disrupting Data in Qualitative Inquiry: Entanglements with the Post-Critical and Post-Anthropocentric* (2017), and co-editor of *Intra-Public Intellectualism: Critical Qualitative Inquiry in the Academy* (2021).

Candace R. Kuby, University of Missouri

Candace R. Kuby is Associate Professor of Learning, Teaching and Curriculum at the University of Missouri, serving as the Department Chair and the Director of Qualitative Inquiry. Dr. Kuby's research interests are: (1) the coming-to-be of literacies when young children work with artistic and digital tools, and (2) approaches to and pedagogies of qualitative inquiry when thinking with post-structural and posthuman philosophies. She has authored several books and numerous journal articles, e.g., *Qualitative Inquiry* and *International Journal of Qualitative Studies in Education*.

Jia Liang, Kansas State University

Jia "Grace" Liang is Assistant Professor of Educational Leadership at Kansas State University. Beside teaching courses in educational leadership, she also engages her scholarship, using qualitative inquiry, and teaches qualitative research methods courses. Her research interests focus on school leadership, teacher leadership, mentoring, diversity and equity in STEM, social justice and equity for women and other minoritized populations, and community engagement. Jia holds a PhD from the University of Georgia in educational administration and policy.

Lauren Mark, Arizona State University

Lauren Mark is a PhD candidate in the Hugh Downs School of Human Communication at Arizona State University. Her work investigates intercultural phenomena, meaning, communication, affect, and relationality within contexts of cultural tension and conflict. She is currently comparing Asian, U.S. American, and global narrations of the COVID-19 pandemic, exploring and developing an Asia-centric methodological approach to the study of public and online discourses. Her work has been published in the

Journal of Intercultural Communication Research, Qualitative Inquiry, Taboo: The Journal of Culture and Education, Culture and Brain, and *Capacious: Journal for Emerging Affect Inquiry.*

Travis M. Marn, Southern Connecticut State University

Travis M. Marn is Assistant Professor of Educational Psychology and Research Methods at Southern Connecticut State University's Curriculum and Learning Department. His scholarship focuses on philosophically informed empirical accounts of performative identities, "Black-White" biracialism, employing new materialism and posthumanism in psychology research, and developing qualitative and post-qualitative methods and methodologies. His work is published in journals such as *Qualitative Inquiry, The Qualitative Report, International Review of Qualitative Research, International Journal of Qualitative Studies in Education,* and *Educational Policy Analysis Archives.*

Susan Naomi Nordstrom, University of Memphis

Susan Naomi Nordstrom is Associate Professor of Educational Research (Qualitative Research Methodology) in the Department of Counseling, Educational Psychology and Research at the University of Memphis. She earned her PhD from the University of Georgia in 2011. She has published about post-qualitative methodologies in leading qualitative research journals such as *Qualitative Inquiry* and *Cultural Studies ↔ Critical Methodologies.* She has lectured and/or given workshops across the United States and in Finland, Stockholm, London, Brisbane, Sydney, and Oslo.

Boden Robertson, University of Alabama

Boden Robertson is a graduate student at the University of Alabama, where he is pursuing a PhD in educational research, specializing in qualitative research. Boden also works at the University of Alabama's Research Assistance Center where he supports students and faculty of diverse disciplines with their research endeavors. Boden's research looks at the experiences of graduate students teaching in a university context.

Stephanie Anne Shelton, University of Alabama

Stephanie Anne Shelton is Assistant Professor of Qualitative Research at the University of Alabama and Co-Coordinator of the Qualitative Research Graduate Certificate Program. Her research, typically interview- or focus group-based, often examines the intersections of genders, sexualities, race, and class in educational settings and the choices that researchers make when generating and analyzing data. Her publications have appeared in journals such as *GLQ: A Journal of Lesbian and Gay Studies*, the *Journal of Lesbian Studies*, the *International Journal of Qualitative Studies in Education*, *Qualitative Inquiry*, and *Qualitative Research Journal*.

Brenda Sifuentez, Lewis and Clark College

Brenda Sifuentez is Associate Professor of Student Affairs Administration and Higher Education at Lewis and Clark College. Her research interests include Hispanic-serving institutions in the Pacific Northwest, the role of space and place, activism in higher education, and organizational change. Sifuentez earned a master of arts in higher education administration and leadership at CSU-Fresno and her PhD in higher education at the University of Denver. She is a graduate of the University of Oregon.

Carol A. Taylor, University of Bath

Carol A. Taylor is Professor of Higher Education and Gender, University of Bath, UK. Carol's research utilizes feminist, new materialist, and posthumanist theories and methodologies to explore gendered inequalities, spatial practices, and staff and students' participation in a range of higher educational sites. She is interested in ethically oriented creative pedagogies, and theory-praxis relations are central to her work. Carol's latest books are: Taylor, Hughes, and Ulmer (Eds.) *Transdisciplinary Feminist Research: Innovations in Theory, Method and Practice* (Routledge), and Taylor and Bayley (Eds.) *Posthumanism and Higher Education: Reimagining Pedagogy, Practice and Research* (Palgrave Macmillan). Carol is Co-Editor of the journal *Gender and Education* and Associate Editor of *Critical Studies in Teaching and Learning*.

María Migueliz Valcarlos, University of South Florida

María Migueliz Valcarlos is a doctoral candidate in curriculum and instruction with a concentration in interdisciplinary education at the University of South Florida. Her research interests lie in the intersection of anti-oppressive approaches to curriculum and pedagogy with culture, language, and technology, particularly the ways they shape being and knowing in education.

Mariia Vitrukh, Arizona State University

Mariia Vitrukh is a PhD student in the Mary Lou Fulton Teachers College at Arizona State University. Her work is grounded in qualitative research and education in the context of war and forced migration. Blending arts-based research, body work literature, and practices, she explores embodied learning experiences, mapping inconvenient and often-forgotten ordeals of forced migration, tracing and collecting live accounts and archival data.

Timothy C. Wells, Arizona State University

Timothy C. Wells (PhD, Arizona State University) is Faculty Associate in the Mary Lou Fulton Teachers College at Arizona State University. His work resides in the fields of qualitative inquiry and the foundations of education, bringing a critical interdisciplinary framework to explore how histories, cultures, and philosophies shape modern education and inquiry. His most recent research considers the affective nature of schooling through archival studies into 19th-century teacher pedagogy and student misbehavior. He has published in *Teachers College Press*, *Qualitative Inquiry*, and *Discourse: A Journal of Culture and Education*.

MENTORING MOMENTS CONTRIBUTORS

Rouhollah Aghasaleh, Humboldt State University

Rouhollah Aghasaleh (PhD, University of Georgia) is Assistant Professor in the School of Education at Humboldt State University. His scholarship lays on an intersection of critical middle grades education, cultural studies of curriculum, and new materialist feminism that addresses the issues of equity and its impact on the education system. Rouhollah's scholarly work has been featured in the *Journal of African American Studies, Journal of Curriculum Theorizing, Curriculum and Teaching Dialogue, Journal of Research in Science Teaching, Activist Science and Technology Education*, and *Reconceptualizing Educational Research Methodology*.

Quintin R. Bostic II, Georgia State University

Quintin R. Bostic II is currently a PhD candidate in the Department of Early Childhood and Elementary Education at Georgia State University. His critical research focuses on how ideas of race, racism, and power are communicated through the text and visual imagery in children's picturebooks. He currently serves as Partnership Manager for the Teaching Lab and Antiracism Co-Chair the National Association for Professional Development Schools (NAPDS). Without support from Dr. Stacey French-Lee, his North Star, he would not be the fearless antiracist scholar that he is today.

Carlson H. Coogler, University of Alabama

Carlson H. Coogler is a doctoral student in educational research with a specialization in qualitative methodologies at the University of Alabama. She is interested in interdisciplinary and artful approaches to learning; embodied, new materialist, and more-than-human theories; and exploring methodologies that entangle theory and practice. A National Science Foundation CADRE Fellow from 2018 to 2019, she is especially interested in researching environmental/ecological ethics, practices, and literacies. She is thankful for all those who have become mentor/mentee with her, especially and including Dr. Kelly W. Guyotte, Dr. Maureen A. Flint, and Dr. Stephanie Anne Shelton.

Nikki Fairchild, University of Portsmouth

Nikki Fairchild is Senior Lecturer at the University of Portsmouth. Her PhD, research, and publications enact posthumanist theorizing to extend conceptualizations of professional practice and more-than-human subjectivities in early childhood. She is also part of a collective, including Carol A. Taylor, Mirka Koro, Neil Carey, Angelo Benozzo, and Constanse Elmenhorst, that entangles with interdisciplinary ways to enact methodology via undisciplined research-creation experimentation articulated in our conference presentations and publications. She is indebted to the nourishing and supportive mentoring from the collective, they have helped her both personally and professionally and she finds it a joy working with them as part of a collective relationality.

Rebekah R. Gordon, Michigan State University

Rebekah R. Gordon is a doctoral student in the Department of Teacher Education at Michigan State University. Her current research focuses on the lived experiences of transnational language teachers and how they negotiate and work across linguistic, cultural, and political systems. The birth of her daughter during her first year of studies has led Rebekah to embrace her identity as a mother-scholar and contribute to increasing the presence of matrifocal perspectives in academia. She is grateful for the ongoing support and guidance from Sandro, Laura, Janine, Lynn, Peter, Alex, Amelia, and of course, Mark and Ophelia.

Sharon Head, Massachusetts College of Liberal Arts

Dr. Sharon Head is Assistant Professor in Special Education at the Massachusetts College of Liberal Arts. She is a former K-12 special educator whose current work is focused on the preparation of the next generation of teacher-advocates. Dr. Head's research centers on the use of narrative methods to better understand the connection between teacher beliefs and their work in the classroom. Her doctoral work was made possible by the radical listening and encouragement of her mentor at the University of New Mexico, Dr. Susan Copeland.

Rachel K. Killam, University of South Florida

Rachel K. Killam is a doctoral student at the University of South Florida in the Interdisciplinary Education program, studying higher education, industrial and organizational psychology, and compassionate leadership. Killam is researching how higher education organizational systems can enact compassion, allowing our full humanity to flourish in neoliberal and colonialized contexts. She has worked in Student Affairs for 12 years and serves at the University of Tampa Office of Career Services. Her mentoring amalgam: "I chuckle imagining the three of you in the same room together. You are starkly different from one another; however, you are all striking powerhouses, and good humans." #WhoRunTheWorld

Minkyu Kim, St. John's University

Minkyu Kim is currently an English teacher at Stuyvesant High School and a PhD student at St. John's University in New York. He would like to acknowledge and thank his mentors: his family; his colleagues in the Stuyvesant English Department; his curriculum and instruction professors and classmates at St. John's; his cooperating teachers Andrew Ravin and Mark Henderson; and his students, who inspire him every day.

Elliott Kuecker, University of Georgia

Elliott Kuecker is a PhD candidate in educational theory and practice at the University of Georgia and an academic librarian.

Krista S. Mallo, Trinity College of Florida

Dr. Mallo serves as Associate Professor of English at Trinity College of Florida. Her research presents poetic autoethnography as an engagement with other qualitative methodologies. Her work marries poetic inquiry with the social science of teacher education through a lens of autonomous learning from a pragmatist feminist approach. She completed her doctoral studies at the University of South Florida's College of Education, Department of Teaching and Learning during the COVID-19 pandemic. She is forever grateful to Dr. Jennifer R. Wolgemuth for mentoring her towards an understanding of the nature and nurture of qualitative inquiry.

Cristina Valencia Mazzanti, University of Georgia

Cristina Valencia Mazzanti is a doctoral candidate under the direction of Dr. Melissa Freeman at the University of Georgia. Her research focuses on the areas of early childhood, language diversity, and qualitative research, specifically as these relate to the study of young children's experiences with multilingualism and learning. Across her scholarship, Cristina aims to develop multifaceted understandings of the social construction of languages that disrupt sedimented assumptions, while arguing for the need for aesthetical and multimodal manifestation of language as representations of knowledge in classrooms and in research.

Shelly Melchior, University of Alabama

Shelly Melchior is a former middle and high school English teacher whose love of learning led her to return to the classroom. She is a PhD candidate in her final semester in the Instructional Leadership–Social and Cultural Studies program at the University of Alabama.

Jaclyn Kuspiel Murray, Augusta University

Jaclyn Kuspiel Murray is Assistant Professor of Science and Engineering Education at Augusta University in Augusta, Georgia. Her research focuses on developing prospective teachers' knowledge and practice of science, engineering, and STE(A)M disciplines and education. She thanks her qualitative research mentors for preparing her to design and initiate a research agenda. The following list includes Jaclyn's mentors presented in the order in which she met them: Barbara A. Crawford (University of Georgia), Jori N. Hall (University of Georgia), Shanna R. Daly (University of Michigan), Seda McKilligan (Iowa State University), and – certainly not least – Colleen M. Seifert (University of Michigan).

Missy Springsteen-Haupt, Iowa State University

Missy Springsteen-Haupt is a middle school teacher, graduate research assistant, and EdD student at Iowa State University studying teacher leadership and social justice in rural communities. Her poem following Chapter 4, for three voices, represents her insecurities and frustrations (left), external and internal discouragement (middle), and the encouragement of her mentor (right). She is grateful for the generous and compassionate mentorship of Dr. Amanda Jo Cordova at North Dakota State University and Dr. Nicolas Tanchuk at Iowa State University.

Aubrey Uresti, San José State University

Dr. Aubrey Uresti is Assistant Professor of Counselor Education at San José State University. Her research focuses on individual, family, and extended family-level implications for and meaning-making processes of adolescents who experience parental incarceration. Dr. Uresti's background as a K-12 educator, school counselor, and therapist informs her exploration of urban education and school counseling, school-based support, grief and loss, peer victimization, child and adolescent development issues, and lifelong learning for counselors through qualitative interview, image-based research, critical discourse analysis, and ethnography. Ever grateful to her mentors, she carries their wisdom and lessons with her into each new academic adventure.

Ying Wang, Dallas Independent District

Ying Wang received her doctorate from the Curriculum Studies program at Texas Christian University in Fort Worth, Texas. During her doctoral studies, she studied closely with her mentor Dr. M. Francyne Huckaby, whom she considers to be instrumental to both her intellectual growth and her sojourning experiences in the United States. Her research interests include curriculum theorizing, cultural studies, feminist theories, and (post-) qualitative research methodology. Her dissertation explores women's doctoral experiences in education with the nomadic subject theory and the affirmative ethics. Currently she works at Dallas Independent District in Texas as a classroom teacher.

FOREWORD

Some of my earliest encounters with Drs. Kelly W. Guyotte and Jennifer R. Wolgemuth were in the context of their roles as mentors within the qualitative research community. Their collective work within professional organizations (e.g., American Educational Research Associations' Qualitative Research Special Interest Group) is well recognized for providing creative and innovative programming that supports mentors and mentees alike. I certainly have benefited from their insights around mentoring, and many of the qualitative methodology graduate students that I work with have as well. Guyotte and Wolgemuth, for me, have long represented the best of mentoring practices within the field of qualitative research, as they have built long-lasting opportunities for graduate students and emerging scholars to receive caring, thoughtful, and timely mentoring; that is, the type of mentoring that sets one up for pursuing the kind of work that brings them joy. Thus, when they shared with me their newest book project, *Philosophical Mentoring in Qualitative Research: Collaborating and Inquiring Together*, I was not surprised, as I have long looked to them as leaders in the area of mentoring within the field of qualitative research.

Indeed, within the field of qualitative research, the topic of mentoring is one that is pointed to often, consistently heralded and supported, and yet so rarely unpacked and critiqued. In theory, I'm certain we all recognize that there are no blank pages to write up – as de Certeau noted decades ago. Instead, we as scholars are always already adding to, expanding upon, and critiquing work that came long before our time. These works and ideas are often introduced to us by our "first" academic mentors and/or by (un)official mentors that live far outside the confines of academia. In fact, within academia, we are commonly told early in our careers, "Make sure you find *many* helpful mentors" – or we are asked, "Who is/was your mentor?" Mentoring becomes a *thing* – most often assumed to be a *person* – that we are implicitly and explicitly told matters and yet so rarely informed about what this might look like or become.

Fundamentally, being a mentor-mentee is part of the work we do, and while there is a vast body of scholarship around mentoring in higher education, the scholarship specific to mentoring in qualitative research is notably minimal. This is really quite remarkable given that our work – as qualitative methodologists and researchers – is so very particular. Our work frequently takes place at the intersection of multiple disciplines and at times in the midst of (contentious) paradigmatic differences. We are also often asked to teach those who are unfamiliar with and even suspicious of the value of qualitative research methodologies and methods. While qualitative researchers have long worked diligently and carefully to build the field and gain increasing levels of acceptance, much work remains to be done, and many individuals who are new (and even "old") in their role as a qualitative methodologist and research find themselves navigating this work in isolation.

In many ways, this edited volume offers what is long overdue – a detailed, complicated, and layered discussion of what mentoring might *become* (or perhaps already *is*) for us as a field. More particularly, the question this volume explores – *How is mentoring a philosophical practice?* – is one that moves us beyond simplistic discussions of doing mentoring and instead provokes considerations of how mentoring is never a given and always in the making. Mentoring and being mentored have real consequences, and part of coming to understand these consequences is recognizing the importance of unpacking the philosophies that undergird one's mentoring practices. Mentoring, as the authors so beautifully highlight, is many things/beings/doings, and mentors and mentees come in many forms. Just like methodology is always evolving and shifting in relation to the contexts and histories wherein it is wielded, so too is mentoring. Collectively, the chapters in this volume illustrate how positioning mentoring as a philosophical practice might lead – as Marn and Guyotte wrote in their chapter – to a "revised ethical vision of responsibility." This "revised vision" is one that I believe is long overdue in the field of qualitative research and can serve as a *new* way forward. Like in so many other contexts where they have led mentoring initiatives and conversations, Guyotte and Wolgemuth (as well as the contributing authors) point to what mentoring might *become*.

Jessica Nina Lester,
Inquiry Methodology, Indiana University

INTRODUCTION

WHY PHILOSOPHICAL MENTORING IN QUALITATIVE RESEARCH?

Kelly W. Guyotte and Jennifer R. Wolgemuth

We introduce this co-edited book on mentoring at a timely inflection point in our academic careers as qualitative methodologists. In the language of the American academy, Kelly is a recently tenured associate professor and Jenni is a tenured associate professor, eyeing the path to full professorship. Despite the growing years between our own doctoral studies and our current positionings in the academy, we still think of ourselves as much mentors as mentees, as much mentored as we are mentoring. If we can make any claims to success in our academic careers, any assertions that we have the necessary expertise to co-edit this text, it is because of the mentoring relationships that support(ed) and sustain(ed) us. But rather than foundationally and hierarchically "standing on the shoulders of Giants," we approach mentoring as relational and multidirectional, happening (and happenstance) within and between a myriad of bodies. Because there are only so many Giants, and not all make good mentors. Also, because it can really hurt when someone stands on your shoulders.

Our formal and informal mentoring experiences include chairing the American Educational Research Association's Qualitative Research Special Interest Group mentoring committee, working with our own assigned and unassigned faculty mentors, and mentoring and being mentored by new doctoral students, to name a few. These experiences sensitized us to the necessity of mentoring as a practice that sustains and advances the field of qualitative inquiry. We are grateful (and we also worked very hard)

DOI: 10.4324/9781003022558-1

1

to have tenure-track positions *in qualitative research*. We understand qualitative research positions are few and far between, both inside and outside the academy. When positions do become available, they often require additional expertise and skills: qualitative research *and* social foundations, qualitative inquiry *and* early literacy, qualitative *and* mixed methods – and, and, and. What the qualitative inquiry community values does not always align with what is important to other fields, governments, and research organizations. Certainly, longstanding and emergent critical and post-qualitative approaches remain in the margins of most academic and research communities. Yet, they are the approaches we and our students recognize as most innovative and cutting-edge, most likely to challenge oppressive forces and prompt much-needed social change; even as opportunities to pursue them are elusive, require creative storying of job letters and research agendas, and not just a little risk-taking. Being a qualitative researcher, or at least doing qualitative research, can be exhausting affective work – empowering and joyful at one moment, gutting and terrifying at another. It is not enough to say, "I did it. So can you." Mentoring relationships sustain us through the challenging times and soar with us during the exciting ones.

Like qualitative research itself, mentoring relationships and practices are relational, affective, and unpredictable, and they are more than techniques and skills to be mastered. Knowing how often to meet your mentee or how to track mentoring goals is not enough to prepare future qualitative researchers to think deeply about methodology, challenge dogma, creatively advance the field, and inspire and activate social change, all the while maintaining yourself and your work as intelligible to broader academic, research, policy, and practitioner communities. Also, like qualitative research methodologies, mentoring and mentoring relationships are guided and inspired by philosophies, orientations to the whys and hows of interactive (see Shelton et al., Chapter 7; Nordstrom, Chapter 9) and intra-active (see Flint & Cannon, Chapter 1; Taylor, Chapter 2; Sifuentez & Gildersleeve; Chapter 8) inquiry relationships, even if those philosophies are not articulated. Mentoring as a philosophical practice moves beyond technique and recognizes that, in many instances, mentoring happens before we (may or may not) take the time to label it as such. Prompted by our recent experiences with mentoring in qualitative research and our concerns about preparing students for success in the field of qualitative inquiry, we are increasingly asking ourselves, our colleagues, and our students, what guides qualitative mentoring practices? What does "good" qualitative mentoring look like? What vision(s) of qualitative research do we enact in mentoring relationships? And, perhaps most importantly, or at least why we pause to reflect on mentoring as we do in this text, why is mentoring something that now seems so essential to the future of qualitative inquiry? Why and how we mentor reflects and produces our vision for the future of qualitative inquiry, one we view as founded less on Giants' shoulders and more on a shared and sustaining net of ideas, affects, and relations.

Literature on teaching and mentoring

A quick search for literature on mentoring in higher education yields ample guides (e.g. Centering on Mentoring 2006 Presidential Task Force, 2006), books (e.g. Johnson, 2015), and articles (e.g. Cramer & Prentice-Dunn, 2007) often geared toward student mentees intent on establishing "best practices" toward "successful" and "effective" relationships with faculty. Some of this literature is geared toward mentoring in specific disciplines, some more general, some speaking to other types of mentoring such as peer-to-peer and new faculty, and some considering the importance of mentoring for historically underrepresented individuals such as students of color and women. With few exceptions, what we find in a majority of this work is a desire to help foster the traditional relational exchange between hierarchies of power: there is someone who possesses some knowledge and/or experiences and/or connections and/or resources, and someone who desires to cultivate a relationship with said individual in order to connect, advance, become successful, etc. Such hierarchies are notable in various publications that posit friendship as incommensurable with "effective" mentorship, creating an either/or (e.g. Zackal, 2019). Certainly, studies have demonstrated the importance of mentoring on and with various individuals, and these mentorship writings have value to many who seek such information. We are, however, interested in taking a different approach to the topic of mentorship. We are concerned with mentoring in qualitative inquiry; yet, we are also interested in what might happen when we disrupt the hierarchies, the traditions, the instrumentalist give-and-take, to consider mentoring differently. This book is the result of our desire for new, relational visions of mentoring in qualitative inquiry that both connect with, and deviate from, the existing literature.

As faculty members who teach/practice qualitative inquiry, we are invested in conversations related to teaching and mentorship in our field. In recent years, we have observed, and even generated, an increasing interest in/for mentoring in qualitative research. We presume the recent interest in mentoring stems in part from a longer tradition of emphasizing teaching inquiry in classroom spaces; the curriculum and pedagogy of qualitative research. For more than a decade, qualitative handbooks (e.g., *SAGE Handbooks of Qualitative Research*) have included chapters on teaching qualitative research; books have been written about teaching qualitative research (e.g., Kuby & Christ, 2019; Swaminathan & Mullvihill, 2018); and prominent journals have dedicated entire special issues to the topic of teaching qualitative research (e.g., *IRQR*, volume 11, issue 3). This emphasis elevates important and even necessary conversations about how and what teaching and learning of qualitative research should entail that reflect perspectives on the field and conduct of qualitative inquiry. To be sure, it is often within inquiry classrooms that learners become introduced to qualitative inquiry and come to question, interrogate, and wrestle with qualitative inquiry as a practice and a possibility for their own research studies.

Even as we are pleased to see conversations of teaching qualitative inquiry imbue conferences, books, and journals, learning to be a qualitative researcher often requires more than formal class and office hour time with professors, more than supervision and collaboration on research projects, and more than exposure to "effective" teaching. This "more than" is what falls under the auspices of mentoring as a close and enduring relationship that cultivates learning and growth, the becoming of qualitative inquirers. Though our focus is on mentoring, we acknowledge that teaching and mentoring are not mutually exclusive; they are, in fact, entwined and mutually informing. Thus, we do not wish to set up teaching and mentoring as dichotomous or even competing facets of academic relationships. Teaching can and does inform our mentoring practices, and vice versa. Teachers can become mentors, and mentors are often teachers even when relations are never grounded in classroom spaces. Moreso, mentees can become mentors and even teachers, productively disrupting traditional hierarchies (see Nordstrom, Chapter 9).

Because we see mentoring as a productive extension of teaching as well as a practice that sparks different conversations, we have conceptualized this book in the "more than" spaces that teaching literature does not often cover. When we discuss the "more than," we are positing qualitative inquiry mentorship as something different from what is typically discussed in the literature on teaching inquiry, and something different from what general mentoring scholarship addresses. In general, we have wondered what conversations on mentoring make possible that conversations on teaching do not. More specifically, as members of the qualitative research community, we wonder what qualitative inquiry mentoring is/does/produces. Inspired by the wealth of literature on mentoring and recent scholarship on teaching inquiry, we situate this text as one that addresses a different facet of qualitative inquiry practices, speaking to the pedagogies within our mentoring relations. The uniqueness of this text is an invitation to consider the philosophical practices of mentoring, as we will discuss further in the next section.

The philosophical practice of mentoring

We have engaged with and even contributed to the literature on teaching qualitative inquiry; however, we realize that there is little scholarship on the practice of inquiry mentoring beyond the spatial and temporal bounds of the classroom or a given research project. What happens outside of a formal class? In conference spaces? Between conference sessions – in hallways, hotel lobbies, and even bars? Who and what are the mentors that spark and inspire one's thinking of/with inquiry? Mentoring, for us, is something that exceeds – it is not contained by classroom or office walls, nor is it produced through hierarchies, nor only experienced with other human bodies (see Coker et al., Chapter 3). Thus, mentoring speaks to both the formal and informal encounters between myriad bodies, in both formal and informal spaces (see Koro et al., Chapter 10). Mentoring happens within, across, and in-between (see Taylor, Chapter 2). Mentoring is affective. Mentoring is relational, sometimes radically so (see Bhattacharya & Liang, Chapter 6).

In this book, we are interested in mentoring practices in the qualitative community that are activated by relations and the ways in which the engagements of mentor and mentee are cultivated, supported, and sustained. As we have discussed, inquiry mentorship is something that has happened for us/ by us/with us, often with little understanding of how we would be affected and how we would affect (see Kuby, Chapter 5). It is not something we are explicitly taught, nor is it something we set out to do; it is often immanent, happening *with*, a relation in-process, always in the middle. Whether we do it because we must, or because we want to, we often do not realize we are inquiring together until we are swept into the profound flows of relationality, until we are already amidst the currents as mentor/mentee.

Despite the importance of mentoring relationships, particularly in methodology, mentoring is often approached as a learn-as-you-go facet of academic life. Thus, scholars often enter mentoring relations with little guidance or knowledge of what such spaces might become, make possible. With this lack of explicit consideration or formal training on the practice of inquiry mentoring, mentoring is often not thought of as a practice inspiring/inspired by philosophy. It is something we do, yet often with little interrogation. This edited book seeks to make visible, interrupt, reflect, deepen, extend, expand, and many possible others, the methodological mentoring practices within the qualitative community. Our overall aims are: (1) to emphasize mentoring as an incessant and necessary part of academic/inquiry work; and (2) to explore the nuanced ways in which philosophies permeate into and inform our mentoring relationships in qualitative inquiry.

Here, we pause to explain what we mean by mentoring as a philosophical practice. When we speak of mentoring *as* philosophical practice, we do not refer simply to one's philosophy *of* mentoring as what might be reflected in a teaching philosophy statement. For us, the prepositional difference between "of" and "as" matters. *Of* suggests the possibility of distance, a stepping away from and cogitating on previous and future actions. The purpose of a philosophy of teaching or mentoring is communicative and often instrumental (e.g. useful for annual reviews and applying for academic positions), as it explains how we teach/mentor and justifies why we teach/mentor that way. We expect that the teaching or mentoring philosophy is an aspect of philosophical practice, but mentoring *as* philosophical practice, for us, denotes the entire mentoring assemblage and its multiple entry points. Mentoring as philosophical practice could be philosophy in action, an enactment of knowing, being, and valuing. Mentoring as philosophical practice could be relational and emergent – a phenomenon with a life of its own. Mentoring as philosophical practice could be interrogated and critiqued. Thus, we sought contributions that went well beyond descriptions and statements of mentoring practices. We sought contributions that coupled mentoring and philosophy to consider how mentoring produces possibilities for knowing and being in the world, even as it is part of their ongoing becoming.

As editors, we invited contributors to consider the ways that inquiry mentoring becomes a practice of philosophy, and perhaps even when it does not/is not. Through their relationality, mentoring practices can disrupt and push against hierarchies, boundaries, normativity, authority, power – and, and, and (see Marn & Guyotte, Chapter 4). How we engage in our mentoring relationships between what is said/unsaid, done/undone, thought/unthought, often spills from our philosophical beliefs and positionings. Here, we endeavor to open these spaces and think intentionally and explicitly about mentoring in qualitative inquiry, emphasizing a practice that is often de-emphasized. This book, then, inquires into the phenomenon of mentoring and asked contributors representing varied philosophical and methodological positions to consider, "How is qualitative inquiry mentoring a philosophical practice?"

The chapters

The contributors of this volume are qualitative researchers, professors of qualitative methodology, graduate students, early career faculty – all of whom value the practice of mentorship, all of whom value the practice of mentorship with philosophy. It was important for us to include those in our qualitative community at various career phases, in various roles, and, in particular, to include contributions with both graduate students and faculty as lead authors. In short, we did not want to simply include writing about effective inquiry mentorship; we wanted mentorship to permeate through the writing/thinking/doing of our contributors. Just as we were intentional with the varied relations represented in this book, it bears mentioning that this text highlights scholars representing post-qualitative, critical, and (de-) colonizing philosophical positions, since, from our vantage point, emerging scholars find themselves yearning for connections, for mentors, who can help them navigate the complexities of these particular philosophical spaces.

For us, philosophy is what and how we live. It is also what and how we inquire. In the chapters that follow, our esteemed contributors have wondered with us about what it means to consider qualitative inquiry mentoring as a philosophical practice, and what such practices make possible. Inspired by philosophers and philosophical concepts, the chapter contributions put mentoring in relation to their philosophical/methodological positionings, just as they/we think of mentoring as a relational practice itself. Thus, we all come (differently) to philosophical mentoring in qualitative inquiry as a process of inquiring together.

Our chapters begin with Maureen A. Flint and Susan O. Cannon's contribution, "Becoming Feminist Swarm: Inquiring Mentorship Methodologically Together." Thinking with feminist scholars such as Rosi Braidotti, Karen Barad, Sara Ahmed, and Donna Haraway, among others, Flint and Cannon story their mentorship intra-actions as a swarming, a way of making kin in the academic world. The chapter is thoughtfully multilayered, moving in and out of emails to one another, reflecting together on important doctoral moments, carefully stitching together the genesis of their mentoring relations as they move between then (student) and now (faculty). Readers are invited to become part of their feminist swarm through embracing relational and intentional practices of making kin in unbounded, evocative, and ethical ways. Philosophical mentoring, for Flint and Cannon, becomes a "mentor-kin-swarm" in which qualitative researchers learn to work (in/with) the tensions and, together, live well.

Also drawing from feminism, Chapter 2 immerses the reader in Carol Taylor's feminist materialism musings entitled "Flipping Mentoring: Feminist Materialist Praxis as Quiet Activism." Taylor re-turns to her earlier writing on mentoring with an enlivened sensibility made possible by Barad, Massumi, Haraway, and others, focusing on how the materialities of mentoring "flip it into a praxis" of feminist activism that quietly corrodes the academy's neoliberal structures and practices. Taylor's chapter is richly theorized. She takes careful time to bring philosophy to bear not just on the praxis of mentoring and the resistances that quiet feminist activism opens up, but also her process of re-turning to mentoring, which brought her the lively concept of flipping mentoring. While sharing emails, photographs, and researcher journal entries collected from a (prior) mentoring relationship, Taylor enjoins the reader in her feminist material analysis of relational flips in mentoring: care, muddle, and time. She invites the reader to rethink with her, to consider these flips as affective agents of response-ability that open "to an intensive in-between" where living, politics, matters become differently possible.

Chapter 3 is a collaboratively and creatively written exploration by individuals brought together in a qualitative research course – Jonathan M. Coker, Samantha Haraf, María Migueliz Valcarlos, Scottie Basham, Diane Austin, Dionne Davis, Anna Gonzalez, and Jennifer R. Wolgemuth – entitled

"Kinning and Composting: Mentorship/t in Post Qualitative Research." This chapter is a challenge to humanist, hierarchical, dichotomized relationships between mentor and mentee. Donna Haraway's work on composting inspires a conceptualization of mentoring as "mentorshit," which takes seriously the role of nonhuman others (e.g., butterflies, cats, mold, urine) who all become compost as mentors in the author's respective journeys in/with/ around qualitative inquiry. Through deconstructing conventional mentoring, the authors' layered narratives connect with the philosophical work of not just Haraway, but Karen Barad and Erin Manning. The mentoring kinship therein rethinks and diversifies how we practice and open up mentorship/t in qualitative research moving forward.

From mentorship/t to ruin, Chapter 4 brings together Travis M. Marn and Kelly W. Guyotte as they envision and explore the practice of "Ruinous Mentorship." With an affirmative position, these authors inquire about what it means to work in the humanist ruins of mentorship, how we are ruined by philosophy, and what such ruin produces, makes possible, in our mentorship practices. Marn and Guyotte entwine conversations about ruinous encounters and events along with a conceptualization of ruinous mentorship in qualitative inquiry, one that thinks with Rosi Braidotti, Judith Butler, and contemporary qualitative scholars and colleagues. Even through ruin, the authors find themselves sustained and even optimistic about the possibilities in ruinous qualitative mentoring.

In Chapter 5, Candace R. Kuby's contribution "Unfinished: (Post-)Philosophically Informed Mentoring and Relational Ethics" slows down and engages with mentorship as a practice of ongoing and expanding relationships. Focusing on ethical relationality, Kuby puts mentorship in conversation with two texts and wonders about the philosophies that inspire her mentoring *with* other bodies – in particular, her experiences with her own academic mentor as well as with students that consider(ed) her a mentor. As she thinks with philosophical concepts like be(com)ing and (unanticipated) resistance, Kuby explores mentoring as both a mundane and lively ontological production – one that is relationally worldmaking.

From ontology to culturally situated ontoepistemologies, Chapter 6 brings together Kakali Bhattacharya and Jia Liang in a work entitled "Mentoring as Radical Interconnectivity and Love: Post-Oppositional and De/colonial Approaches to Qualitative Research Mentoring in Higher Education." Here, Bhattacharya and Liang write from the perspectives of women of color with post-oppositionality using critical and de/colonial philosophical frames. Grounded in their individual (and collective) experiences navigating academe, these authors conceptualize and invite the reader to consider four mentoring practices that cultivate radical interconnectivity and love. Bhattacharya and Liang's culturally situated approach beautifully elevates vulnerability as it envisions a radical and relational future for mentoring in qualitative inquiry.

Alongside love and vulnerability is tenderness, introduced in Chapter 7 by Stephanie Anne Shelton, April M. Jones, Kelsey H. Guy, and Boden Robertson, "The Importance of Relationships in Mentorship and Methodological Identities." The authors – a professor and students – weave a reflective story of their seemingly unlikely coming to qualitative inquiry and the monumental importance of mentoring relationships in their individual and collective journeys as #qualleagues. Drawing on Thompson's pedagogy of tenderness, their philosophical mentoring is caring and careful, focused deeply on the sacredness of relationships and through which methodological identities are nurtured and grown.

Chapter 8 by Brenda Sifuentez and Ryan Evely Gildersleeve "'I Don't Want Feelings. I Want Tacos.': Toward a New Materialist Mentoring Practice" advances an ethic of care unladen by the trappings of conventional humanist feelings. Drawing on new materialist and assemblage theory, Sifuentez and Gildersleeve conceptualize philosophical mentoring as becoming-mentoring – an orientation that is alive to the materiality of mentoring materials and contexts: the smell of tacos, a university bench, the hierarchical structures of university environments and roles. Foregrounding the materiality of mentoring relationships, while foregoing emotional labor, engenders a sense of trust and reliance in process – *becoming-mentoring* – that is also vital to the (unknown, unknowable) ongoing becoming of qualitative inquiry.

What might it look like to mentor in a flat ontology? This question is taken up by Susan Naomi Nordstrom in Chapter 9, entitled "String Figure Mentoring in a Qualitative Inquiry Assemblage." Drawing inspiration from Donna Haraway among others, Nordstrom conceptualizes string figure mentoring as a relational practice in qualitative inquiry – one that disrupts hierarchy, calls upon Gilles Deleuze and Felix Guattari's logic of the *and*, and posits mentor as "string figurer" in supple and shifting assemblages. Throughout, Nordstrom considers her own practices in qualitative mentoring and thoughtfully explores the passing and crafting of mentorship, pondering what mentoring might produce, might become, in a flat ontology.

We conclude with Chapter 10, a collaborative and beautifully multilayered work by Mirka Koro, Mariia Vitrukh, Nicole Bowers, Lauren Mark, and Timothy C. Wells called "Mentoring (Maybe) as a Philosophical Event." Here, the authors position mentoring in qualitative inquiry as a series of philosophical encounters – ongoing, relational, and experimental – that extend from their experiences in/with higher education, and with one another. The chapter is an invitation toward mentoring as a philosophical practice in qualitative inquiry, opening up just as the book itself concludes. Along with the authors of this final chapter, we nudge our readers to consider their own mentoring encounters and experiences, while dwelling in the possibilities of what might (be)come.

Mentoring Moments

The COVID-19 pandemic brought forth unprecedented challenges for many academics, including some of our original contributors. As we considered what to do with the (literal) spaces left by those who, for whatever reason, were unable to continue on with their writing projects, we decided to elevate the voices of graduate students and non-tenured early career faculty regarding their mentoring experiences. Thus, we called these contributions "Mentoring Moments" and invited individuals to consider four questions through written, poetic, visual, and/or other creative modalities: (1) What does mentoring make possible? (2) What does mentoring open up? (3) What is the beauty of/in mentoring? (4) When has mentoring become salient to you? Just under 30 individuals responded to our call with submissions, and we were faced with the difficult task of culling and curating Mentoring Moments related to the scope of the call and the book itself. We turned, then, to the contributions that provided unique theoretical or identity perspectives, that pushed creative boundaries, and that explored mentoring experiences through thought-provoking vignettes – all of which spoke in some way to the notion of qualitative mentoring as a philosophical practice.

Situated betwixt and between the chapters discussed in the previous section, 17 Mentoring Moments serve as interludes that connect with, extend, expand, and sometimes even juxtapose conceptions of mentoring brought forth by our esteemed authors. We offer such vibrant moments to further our positioning that mentoring in qualitative inquiry is multiplicitous – there is no *one* common experience amongst mentors/mentees, there is no *one* best practice as mentor/mentee. Instead, you will find a collection of both significant moments and minor gestures that linger with these emerging scholars. We invite you, our reader, to pause with each of these Mentoring Moments as you move through the book, to think about your own lingering moments and the possibilities, openings, beauty, and salience therein.

An(other) invitation

We co-edited this book for those who are searching for inquiry mentorship, who want to take the time to ponder about inquiry mentoring, who had mentors/are mentors/want to become mentors for others in a vast and varied qualitative community. Whether you are new to qualitative inquiry or have been part of this field for many years, whether you teach qualitative inquiry or are an aspiring qualitative researcher, whether you have an extensive philosophical background or are new to philosophy, we invite you to engage with our text where you are. When you encounter the chapters to come, we hope the ideas strike resonance, push, provoke, inspire, and nudge you to consider further the (and your) practices of qualitative inquiry mentorship and/with philosophy. To be sure, there is no "right" way to feel or to engage with what follows.

Our invitation is to enter into the chapters in any way that inspires you. We have organized the text with attention to the flow and connection of ideas, sometimes similar, sometimes disparate. Yet we hope you will not see the chapters as unfolding linearly, but enfolding various quandaries, sparks, and topics that can be engaged in various ways. These writings are in dialogue with each other, just as they have the possibility of engaging dialogue with readers. Regardless of how you enter, we invite you to inquire with us, and all the authors, as we together consider the possibilities of philosophical mentoring in qualitative inquiry.

References

Centering on Mentoring 2006 Presidential Task Force. (2006). *Introduction to mentoring: A guide for mentors and mentees*. www.apa.org/education/grad/intro-mentoring.pdf
Cramer, R. J., & Prentice-Dunn, S. (2007). Caring for the whole person: Guidelines for advancing undergraduate mentorship. *College Student Journal, 41*(4), 771–778.

Johnson, W. B. (2015). *On being a mentor: A guide for higher education faculty* (2nd ed.). Routledge.

Kuby, C. R., & Christ, R. C. (2019). *Speculative pedagogies of qualitative inquiry*. Routledge.

Swaminathan, R., & Mullvihill, T. M. (2018). *Teaching qualitative research: Strategies for emerging scholars*. Guilford.

Zackal, J. (2019, May 13). Why you should develop a mentorship rather than a friendship. *HigherEd360*. www.highered360.com/articles/articleDisplay.cfm?ID=1935

WILLFUL HABIT

Shelly Melchior

She shrinks
and the room expands,
and all the spaces are filled
whiles she watches and wonders.
A ghost presence.

It is willful
and habit.
A *willful habit* to become the least,
believe the least,
Be There and yet Not There.
It is how she moves through space
Filling
yet not full.

She is a gradual disappearing act
who decides this space too, is not for her.
And who will miss this presence that is not a presence. . .
Will she?

 DOI: 10.4324/9781003022558-2

Unsure of who she is
and if belonging is a thing
she can ever truly know
outside of motherhood, daughterhood, and wifehood
until someone notices.

And in the noticing
which is both infinitesimal and vast at the same time
an outline begins to fill
shades and tints and hues
that become brighter in unused places
in these hallowed spaces
where belonging was something that belonged to others

And in that noticing
those meetings
the emails
the ideas
the writings
the coffee shops without coffee but with ideas and notebook pages filled
with back and forth

A voice that believed she had no place other than the one defined for her
given to her
begins to shape and define itself – for herself.
untapped, undiscovered, unmined
until one someone took the time
to notice the outline in the classroom
and to believe

and she was there all along. Just waiting.

1

BECOMING FEMINIST SWARM

INQUIRING MENTORSHIP
METHODOLOGICALLY TOGETHER

Maureen A. Flint and Susan O. Cannon

< < < < > > > >

I am so glad we got a chance to chat briefly at AERA.[1] I would very much be interested in taking up some reading or thinking together. . . . Let me know what you think or if you are interested at all, perhaps I'm being presumptuous.

It is so good to hear from you, and I would be honored to take up some reading or thinking with you – at ICQI and beyond. Do you have a text in mind?

< < < < > > > >

How do you tell the story of a meeting? The point of possibility that became a swarm of friendship, mentorship, writing/reading/thinking togetherness? Perhaps we could start with Maureen reading Susan's article (Cannon, 2018), or Susan joining the mentoring committee of the QR-SIG[2] at AERA, where we first met. Or we could map farther back, to Maureen taking an introductory qualitative research course and applying to the PhD program in educational research after finding kinship and resonance in the intersections between art and qualitative research, a kinship generated through Kelly W. Guyotte's artful pedagogical practices. Maureen,

---

1   AERA is the acronym for the American Educational Research Association.
2   QR-SIG is the acronym for the Qualitative Research Special Interest Group at AERA.

   DOI: 10.4324/9781003022558-3

a PhD student, taking the next class in the qualitative course sequence, where Kelly assigned Susan Nordstrom's (2015) *A Data Assemblage*, which led her, zigzagging through related journal articles, to an article by Susan about transcription (Cannon, 2018). Months later, Kelly forwarding Maureen an email from the QR-SIG at AERA about the annual mentoring conference session, urging her to apply. Maureen attending the session and meeting Susan there, and the moment of recognition sparking between text-theory-mentorship. Following the spark, Susan emailing Maureen, making plans to meet at ICQI[3] a few months later. Meeting, having drinks. Exchanging numbers, plans to meet virtually over the summer. A Zoom meeting, a reading schedule, engaging with *Meeting the Universe Halfway* (Barad, 2007) together over the summer. And then both (separately) being told by our mentors, advisors, dissertation chairs, that we needed to start writing, to focus. No new reading.

So, we moved from reading together to something else. Sending each other snippets, segments, ideas, drafts. Writings, outlines, and sketches that were shared through the foundation of our initial kinship, our initial meeting. Sharing texts that were messy and raw, fragmented and scattered, as we moved through the process of stitching together our dissertations. As we shared writing each week, we began to cast on other stitches. Threads of applying for jobs and on-campus interviews, of teaching and academic partnerships, of life and relationships. These threads bundled up and tangled and twisted, complicating what we thought we knew of the other person based on their mentorship (she is Kelly's student) or their philosophical guide (she reads Barad). Our identities became more complicated to each other, pieces buzzing about, destabilizing the idea of stable selves and our academic identities. In this way, we began to swarm, finding connections and relations, tentative undulations that became a rhythm. Donna Haraway (2016) wrote that "the task is to make kin in lines of inventive connection as a practice of learning to live and die well with each other in a thick present" (p. 1). Dwelling in the thick present of graduate school, dissertating, becoming academics, we cast on threads of connection and relation. Becoming together, making kin, swarming in textual, material, and emotional intra-actions that took us virtually and physically into each other's homes, the homes of friends and relatives, and

---

3   ICQI is the acronym for the International Congress of Qualitative Inquiry.

McDonald's and coffee shops across the country. In our emails to each other, we wrote,

> Thank you for sharing your work with me. I realize that we have not built a lot of trust and only know each other a bit, so I wanted to give feedback with care. I hope that it makes sense, and you receive my comments as offerings to take or leave – especially given the time frame.
>
> I also enjoyed reading your paper, and am looking forward to your reading of/with the data – I made a few comments through the document, though given word count, some are more just musings of lines that your writing produced/thoughts that a longer paper could elaborate on . . .

Practicing an "ongoing alertness to other-ness-in-relation" (Haraway, 2016, p. 141), we think our ongoing qualitative mentorship with Ahmed (2017) as feminist homework, "mak[ing] everything into something that is questionable" (p. 2). Our swarming, then, is feminist in that we centered an ethics of care and relationality in our becomings-with each other, even as we question, critique, and trouble the taken-for-granted aspects of mentorship. The swarm indicates the attention to the micro and macro aspects of relationality and interconnectedness. As Ahmed (2017) wrote,

> [feminism is about] asking ethical questions about how to live better in an unjust and unequal world . . . [finding] ways to support those who are not supported or are less supported by social systems; how to keep coming up against histories that have become concrete.
>
> (p. 1)

Throughout this chapter, we evoke multiple feminist swarms, multiple thinkings and becomings together. The past swarmings of our dissertating: excerpts of emails and texts sent back and forth, comments inserted in margins of drafts; our present-future swarmings as we think and write together through this chapter and our current academic roles as junior faculty on the tenure track; and the past-present-future swarmings of the scholars, mentors, and colleagues we cite and draw from through this chapter. In this way, our swarm picks up and lingers in multiple spaces and times simultaneously, always becoming, buzzing with mattering in both unexpected and familiar places and spaces.

Drawing on Haraway (2016), we see the swarm as a way of making kin, and thinking with Barad (2007), we recognize the co-constitution of our academics selves with and through the swarm. Braidotti brings affirmative connections within and beyond the swarm. Collectively, we wonder how much we could undulate to make room to become differently in the academy, sharing texts and theories, "Playing games of string figures . . . dropping threads and failing but sometimes finding something that works, something consequential and maybe even beautiful, that wasn't there before" (Haraway, 2016, p. 10). Thus, we work in this chapter, guided by feminist theory, to expand what forms and bodies mentoring can take, leaning into each other, creating a two (four/six/nx2) headed swarm, becoming dissertation/academic/scholar/friend/feminist.

In what follows, we first situate the concept of mentorship produced in and by the literature, then move to discuss how we are conceptualizing mentorship(s) and mentor(s) differently. Then, we move through examples of our swarming. Throughout, we weave in excerpts from emails sent back and forth throughout the year in which we both worked on our dissertations, "data" that we feel is illustrative of our kinmaking, the string figures we composed together, our (sometimes tentative, sometimes exuberant) connections and becomings. Throughout, we conceptualize this as a feminist swarming in that we were guided not by a set of rules or moral imperatives, but by ethical pulls and twists, an impetus to live well together. And, this becoming-with, this togetherness, was not limited to just the two of us but spiraled out to include the texts and theories we read, friends, family, and mentors, our universities and the ideas of the academy itself, including our citation practices. As Ahmed (2017) wrote, "Citation is feminist memory. Citation is how we acknowledge our debt to those who came before; those who helped us find our way when the way was obscured because we deviated from the paths we were told to follow" (pp. 15–16). Thus, our reference list becomes another swarming, filled with the buzzing of feminist theorists and colleagues and friends that have become with us, challenging us to live and think differently.

Throughout this chapter, we offer these examples of our swarming not as a map to trace, an outline to fill in – "this is what mentorship should be" – but rather as a practice of telling stories, of world building, "relaying connections that matter, of telling stories in hand upon hand, digit upon digit, attachment site upon attachment site, to craft conditions for finite flourishing on terra, on earth" (Haraway, 2016, p. 10). Thus, the telling of our swarming is a methodological practice, a practice of drawing string figures, wondering about other forms and versions of mentorship in qualitative inquiry. Our examples offer opportunities for plugging in, for tying on, for clicking in, and for swarming with us. We invite you to think/become mentorship differently with us, to become part of the swarm.

Mentorship/making kin/making swarm

Research on doctoral students emphasizes the importance of mentoring relationships between faculty members and students for degree completion, acculturation, and successful promotion in the academy (Ali & Coate, 2013; Dohm & Cummings, 2002; Grant & Simmons, 2008; VanTuyle & Watkins, 2010). In particular, research on mentorship highlights the importance of mentored relationships along multiple identity axes, emphasizing the ways in which women, and women of color in particular, experience heightened marginalization as a result of the magnifying effects of sexism and racism (Grant, 2010; Magano, 2011; Pope & Edwards, 2016). For example, Dohm and Cummings (2002) found that women graduate students who had been mentored were more likely to do research, and VanTuyle and Watkins (2010) found that mentorship increased professional productivity for women after graduation. Beyond simply improving rates of completion, mentorship is often cited as an intervention, a way to support doctoral students as they navigate the hegemonic and often patriarchal structures of doctoral studies and academia (Acker & Dillabough, 2007; Aiston & Jung, 2015; Grant & Knowles, 2010).

Even as literature on mentorship emphasizes the importance of relationships, the literature frequently positions mentorship as a linear and hierarchical relationship between faculty and student, one of knower to known. For example, VanTuyle and Watkins (2010) described "stages and step-by-step processes that were utilized by both the mentor and mentee to obtain [a] highly sought after position at one of the fastest growing programs for educational leadership" (p. 209). Similarly, Grant and Simmons (2008) developed an eight-step model for mentoring

African American women at PWIs that mapped a linear "progression that begins from doctoral entrance and graduation, acquisition of the university faculty appointment, and to promotion within the professoriate" (p. 512). In this way, much of the literature positions mentorship as an intervention to be applied, steps to be followed to achieve a particular outcome.

However, other research, often scholarship that makes connections to feminist and intersectional frameworks, emphasizes the collaborative, multilayered, and multidirectional ways in which women doctoral students (and female faculty members) collaborate and develop sustaining, supportive, and generative relationships as a means to persist and succeed in academia (Brooks et al., 2020; Flint & Guyotte, 2019; Pope & Edwards, 2016). For example, Shelton and colleagues (2019) described the ways in which women doctoral students participating in a focus group "(re)birthed themselves on their terms, for themselves, even while still/always negotiating the gender norms that continued to work to crush and mold them, in academia and beyond" (p. 138). Importantly, in Shelton and colleagues' descriptions of women doctoral students (re)constructing and (re)constituting what it meant to be(come) as women in the academy, the connections and community formations were not linear or hierarchical, binary, or transactional. Rather, the women described multidirectional, complex connections between and among other students, faculty, texts, family, friends, experiences, coursework, and beyond. Shelton and colleagues described these generative connections as becoming (wo)monstrous, women doctoral students "imagining new ways of being, while wreaking havoc on hegemony and hierarchy" (p. 112).

Like the women in Shelton and colleagues' focus groups, our telling of our mentoring is not a one-way narrative between us, but collects together advisors, senior scholars and theorists, texts, other graduate students, and our writings. In other words, as Shelton and Melchior (2020) described in their mentored relationship between an assistant professor and graduate student, "ours is no Genesis story, with a something that emerges from a nothing; our stories are ones that have always been jumbled in the chaos of others" (p. 53). Even as we invite you to plug into our swarming, the immanent feminist buzz of care and kinmaking and relationality always already extends behind, through, and in front of us.

Formative flights and stutters

We picked each other out of the hive. Though we were working within existing formal mentoring relationships in the form of our doctoral committees, advisors, and AERA QR-SIG mentoring sessions and office hours, we turned toward each other through the dissertation process. This is not to say that we did not send drafts to committee members or discuss theory with them. However, we gave each other the rough drafts, the incomplete thoughts, the data arrangements that we were not sure made sense. These exchanges began tentatively, and the writing we exchanged at the beginning was certainly more polished, our emails to each other apologizing, situating, orienting the other to our thoughts and data and theory. By the end, we were sending writing that we would not have sent to anyone else – writing that we thought of as too raw or unfinished. It was in these exchanges that we both felt free to experiment, question, and critique each other's ideas, to offer and ask of the "and yet" (Taylor, 2019).

> Thanks again for reading this. Whew. It needs work, but you reading it helped me get back in there and cut some of the fat. Reusing writing is not all it's cracked up to be!
> I'm still stuttering and stumbling a bit with trying to move through it all – but I think (hope) there's enough of a start so that you can provide some feedback on if the map I'm making makes sense.
> This is a hot mess . . . there are places where I just copied in quotes and notes of what I am thinking. And the organization is pretty terrible re: flow, etc. Read what you can . . . And we'll chat.
> Rambling comments and thoughts throughout!
> I'm here but I don't think you can hear me
> I'm (as always) struggling with multiple rabbit holes, figuring out how to hold a thread and keep with it. I'm trying to talk about what I did because it's necessary I think, but and do it in ways that are open to complication – right now, this is near the beginning of my dissertation document (before literature and theory . . .)
> I don't know, perhaps writing about it was actually more useful just as a thought exercise.

There was something about the hesitancy with which the other would introduce the writing that let the other know that we could play here. This was not a mentoring in the sense that we were demonstrating or saying to the other, "it would be better this way"; instead, we offered inventive lines of connection to the other. Offerings of what might it look/sound/feel like to write this way or that way, to think that thought or that, to invoke that theory, follow a line, a scent. We offered metaphors, or extended them; we went off track at times. Through reading one another's work, we became kin, made swarm, together. Through texts, emails, feedback, and comments sent back and forth over the course of weeks and months, we learned different ways of writing, of disserting, and of being in the academy. Developing an expanding swarm, moving from center to periphery.

All-but-desiring-dissertations

Comprehensive (or qualifying) exams, prospectus (or proposal), and the dissertation are all pedagogical and disciplining structures in the academy. We become with and through them (Cannon, 2020; Guttorm, 2016). Dissertations produced before us are mentor texts that we are asked to read and review and learn from. Whole books are dedicated to the journey that is the dissertation with trajectories and travel tips laid out for easy (or easier) navigation. A recipe to follow, a journey to replicate. Coursework, comprehensive exams, prospectus, dissertation. Each point in the line toward completion is both a test and a milestone of achievement, even as that line is anything but linear, anything but replicable. In particular, the lines between an approved prospectus and dissertation are tenuous and tangled. On the one hand, we know *enough* to have our research plan approved; on the other hand, we are producing texts and a document that will determine both our legitimacy and our place in the field. Legitimate, enough. Enough signatures to pass a defense; enough to have a sheet of paper signed, sealed, and delivered to our respective graduate school administration; enough to graduate, to walk across the stage, velvet hood slipped over a bowed head. Yet, often, in the period between proposal and defense, it is the other enoughs that linger, as Maureen wondered elsewhere, "Did I represent the students who took part in [my] study adequately, richly, evocatively, enough? Did my research do enough in the world? Did I say enough to make a difference?" (Flint, 2020, p. 3).

The lingering of these multiple enoughs that arise in the period following a proposal defense can be paralyzing. A space of uncertainty where, contradictorily, we are supposed to "at least appear certain in the uncertainty about the subject so they [we] can prove movement along the path" (Myers et al., 2017, p. 313). Further, research on doctoral student completion and attrition has highlighted the time after proposal as "high-risk" (Kelley & Salisbury-Glennon, 2016; Liao et al., 2009, p. 482). As a period of "high-risk" time, the slice of time-space known as "ABD," or all-but-dissertation, is often the period where research on doctoral student retention and completion emphasizes mentorship and community as being particularly influential (Liao et al., 2009; Locke & Boyle, 2016). ABD, a time-space continuum characterized by the ending of coursework, comprehensive exams, and the successful defense of a proposal, can extend multiple years and is often characterized by "numerous emotional and developmental conflicts" (Blum, 2010, p. 74). As Blum (2010) further noted, all-but-dissertation can become a way of living and being. Many doctoral students do not move past the ABD phase, caught in the liminal space between proposal and completion. As recent ABD candidates at the time of our meeting, the anxieties of rates of completion, of doing enough, echoed through our conversations with each other. Included in these anxieties were the desires to do well, to make our mentors and faculty proud, to mark ourselves as deserving of careers in academia.

These desires to do well, to do enough, to produce successful dissertations, are not limited to doctoral students. The formal structured mentoring relationships in academia, however relational and ethical and caring, are always embedded with power dynamics both institutional and personal, the desire to please, the desire to protect, the need to push, the feeling of responsibility. As Hughes and Vagle (2014) wrote, "faculty members *find themselves in* what I would call *pedagogical ambivalence*, between wanting to be responsive to the individual student's desires and wanting to be answerable . . . to dissertation traditions" (p. 254, emphasis our own). Our successful production of a dissertation, then, not only becomes an indication of our success as scholars but also is tied to our advisor's success or failure as mentors (Myers et al., 2017). There are pressures on faculty to graduate doctoral students within a certain amount of time, to produce scholars who are legible in the field, to produce students who can acquire tenure-track jobs. Both of our committees worked the balance between attention to the field for which we were being produced and the relational and ethical responsibilities to our complexity.

Haraway (2018) described kinmaking as "taking the risk of becoming-with new kinds of person-making, generative and experimental categories of kindred, other sorts of 'we,' other sorts of 'selves'" (p. 102). Our relationship, rather than being in opposition to the formal mentoring provided by our committees, came from a desire to make kin, become together. In other words, we did not begin our relationship because of the literature documenting the need for mentorship as a means to be more productive, or because our committees advised us to. There was an intensity that we felt, a kinship, a swarming of our ideas. As we talked, we could trace lines throughout our readings and writings and thinkings that were intricate and complex; there were multiple crossings and connections that we wanted to explore. In the knottings of our relationship there was a kinship, a shared tangle of related influences and relations and companions in making our lives together across different spaces and times.

This just popped into my head, but these anecdotes/memories/narratives you are using to point to the way that truth is constructed/produced are evocative – and your tone of writing/cadence is different in them as well – I almost feel like they could be interludes or brief pauses – might help the reader (hah, we are always so concerned about them) to move between the two styles of writing I'm seeing . . .

Thinking in this section (because I've been returning to/revising it in my own document) – how do I write up my study? How do I tell what I did/describe my methods in a way that feels consistent with the rest of my writing . . .

Wondering if you might have time to read what will be my chapter 1 and give a bit of feedback. Really just one read through, not a super deep reading . . . I am wondering when/if you read it what your takeaway is or if there are ways that I can strengthen the message.

Lately I've been thinking about this phrasing in my own writing, with black and brown bodies/black bodies – and I've shifted to using lives (and not bodies) or lives and bodies – I don't know, I'm still thinking with it, but it feels like for me that to say "bodies" (even as it acknowledges the role/impact/importance that race places in those students moving through the world), is also to do some violence, to objectify black and brown lives down to bodies – I don't know, that's not an answer, just something I'm struggling with now . . .

Is this something that you have used before? I like it, I wonder what it does/means/signifies (even as maybe you don't want me to know . . .). I also wonder about becoming excess – creating new arms/legs appendages to make new cracks, to squeeze through multiple cracks

Many-headed kinships

Separately, we had each built nests of mentor texts, theorists, and philosophers that we turned to, talked with, and invoked in our writing and conversations. Though we were writing and thinking in different fields (mathematics education, higher education) with different theorists (Barad, 2007; Braidotti, 2006, 2011) our nests had overlap, beginning with the shared reading of Barad's (2007) *Meeting the Universe Halfway* the summer after we first met. Before meeting, we had both felt isolated in the nest of mentor texts we had composed. As we began to work and write and think together, our nests of texts overlapped, becoming a hive. As Brooks and colleagues (2020) wrote of their collective biography, "we hear each other. We write with/to/for/from each other. We tangle and form tangles of stories and experiences, interwoven and tight" (p. 297). Sending feedback to each other, our texts converged and differed as we developed an ongoing alertness to otherness in relation.

For example, Maureen (2019) cited N.K. Jemisin (2018), writing in a draft of her concluding chapters that she wanted to "spin the futures I want to see . . . [and] talk back at the classics of the genre" (p. 2). In the middle of reading Maureen's writing to give feedback on the draft, Susan ordered the book, desiring more of that language, that thinking, that feeling she got from reading that quote. Our nests of philosophical mentors proliferated as we traded and collected mentor texts (Ahmed, 2017; Barad, 2007; Haraway, 2016), gathering a collection of metaphors that grew and expanded out and became threads and currents in our dissertations (Cannon, 2019; Flint, 2019).

Susan would be on her morning walk thinking about Maureen's writing about the "spacetimematterings" of a confederate monument, rewriting bits in her head. Maureen, sitting at her sewing table, would find herself thinking with Susan's diffractive readings and poetic reconfigurations, finding bits of Susan's turns of phrases and ways of writing in her own work. We lived differently through reading each other's dissertations, cultivating curiosity and invention, lingering in each other's work and thinking. Visiting with/in each other's studies. Haraway (2016) described visiting as a practice of "find[ing] others actively interesting . . . ask[ing] questions that one's interlocutors truly find interesting, cultivat[ing] the wild virtue of curiosity, retune[ing] one's ability to sense and respond" (p. 127). As we visited, we traded texts and chapters, rewriting, deleting, rearranging. We wrote ourselves into one another's texts, affirming, challenging, suggesting, swarming. Visiting in this way cultivated the possibility for "something *interesting* to happen . . . [through] letting those one visits intra-actively shape what occurs. They are not who/what we expected to visit, and we are not who/what were anticipated either" (Haraway, 2016, p. 127).

Our visits were not only relational, but temporal. Time spent each week discussing our readings, time spent reading and commenting along the sides of the others' work. Time as a disciplinary device, keeping us on track, waiting for the other's writing each week. Hancock and Fontanella-Nothom (2020) wrote of their becomings in an advanced qualitative research class and described how they embraced the tensions of time and that "through collaboration we continuously produce[d] sustainment 'beyond' the classroom amid flux . . . allow[ing] uncertainty and curiosity to percolate, to produce new concepts, which might sustain and extend our nourishing pedagogical experiences" (p. 86). Time also morphed and shifted in non-linear ways, creating time and space to experiment, to play. There was a rhythm to our visits, hummings among the swarm, returns and refrains as we grappled with theory and areas of stuckness.

> I frankly do not have something on diffraction/apparatus that is finished enough to send to you
> > I'm sending this now just to have it to you – I'm looking for streamlining, and I don't think there are any headings or

something at the moment, and I get kind of lost in it, so any
feedback on organization is also helpful!
If you are sick of this piece like I am – no worries . . . Looking
forward to chatting tomorrow.
Ok . . . so this is long and still needs lots of work. Good
thing I have six more days. I did not place the versions side by
side as we had discussed because it was too messy on the page.
It's long. Do what you can . . .

Aside from keeping us "on track" in finishing our dissertations, trading texts strengthened our resolve to be able to write our dissertations differently. Relative to our institutions and fields, we were both moving outside of the lines of the conventional, five-chapter or three-article format of dissertations. Our feedback to each other zigzagged between writing and thinking together, and it was not always about the writing itself, but about being/becoming in relation. Affirming the moves and spirals and paths the other was taking. Maureen, spiraling beyond five chapters. Susan, reading the same data over and again, thinking across fields. By making kin (we were both doing something different, taking risks in methodology and theory) we explored how we might become scholars differently together. We found, just as Van Cleave and Bridges-Rhoads (2013) before us, that "our work together as writing partners repeatedly complicated each of our understandings of the theories that guide[d] our work" (p. 677). Maureen's readings of Braidotti (2006, 2011) complicated Susan's readings of Barad (2007) and vice versa. Lines were drawn and tentatively connected between theories and philosophies and data. We had each read some of the other person's theorist as well, so we brought our understandings/meaning/readings and tossed them back and forth, creating an intermeshed felt of ideas about responsibility and ethics and becoming that multiplied and proliferated.

As we read each other's work, we explicitly talked about the moves we made outside of the normalized boundaries of the dissertation. We wondered, together, how messy we could be, how we could help our readers along the methodological paths we drew, even as we ourselves did not always know where they were going. For example, Susan writing for a mathematics education audience, ultimately decided to write a five-chapter dissertation as a move toward recognizability since the theories and methods she was taking up were already a stretch for the field of mathematics education. We discussed stretching the field and the balance of being simultaneously legible and pushing and pulling at the boundaries of legibility. We wondered how far outside the swarm we could fly. Being outside, too illegible, would leave us cordoned off from the fields that we hoped to shape. At the same time, we worried that if we were not bold enough in our dissertations, in formatting, in content, in composition, that our dissertations as future mentor texts would not open up enough space for those following us to elbow their way into other ways of doing.

We shared and traded advice across our committee members, and advisors, Kelly and Stephanie becoming Susan's mentor in a way (a part of her swarm), Sarah/Teri/David becoming part of Maureen's swarm. We played with each other's ideas, writing into each other's texts, sentences, and paragraphs, taking chunks from feedback, haphazardly written in comment boxes on the side, pasting them into the bodies of our dissertation, writing through them. This practice of writing through and with challenged traditional ideas of authorship, traditional ideas of the dissertation as a solo-authored artifact. Van Cleave and Bridges-Rhoads (2013) problematized citational practices in their dissertations using the construct of "as cited in" to "embrace and account for the fact that our citations of Foucault, and many other philosophers for that matter, are always already produced in writings with one another" (p. 678).

We recognized that our thinking/living/being/writing is always already entangled with all the mortal critters and materials with which we interact; and we noticed that this particular interaction of Maureen/Susan/Barad/Braidotti/dissertation and on and on produced speeds and slownesses that worked for our thinkings/becomings. As Haraway (2016) urged us,

> it matters what matters we use to think other matters with; it matters what stories we tell to tell other stories with; it matters what knots knot knots, what thoughts think thoughts, what descriptions describe descriptions, what ties tie ties. It matters what stories make worlds, what worlds make stories.
>
> (p. 12)

Thinking with Barad's (2007) objectivity, and the concept of accounting for the specific material arrangements of our knowledge-making practices, we each mattered in each other's dissertation. The swarm(s) were complex and entangled, multifaceted and overlapping, and they allowed us to make different ethical movements, connections, and relations. We claimed each other as mattering in the acknowledgment sections of our dissertations. But to trace exactly how we mattered or which lines of text might need to be cited seems an impossible task. We were always cutting together/apart (Barad, 2013), making connections and writing in and toward, and affirming/negating stable author and legible scholar. In a way more than writing *with* each other, or writing *into* each other's dissertations, each of us was mentoring the text of the other. Swarming, mentoring, kinning. We brought our swarm of feminist mentors to each reading and writing, so Susan's edits/suggestions to Maureen's texts were read through Barad (2007) and Haraway (2016), and Maureen's to Susan's were read through Braidotti (2006, 2011) Massey (2005), and Ahmed (2012, 2017). We attempted to mentor each other's text toward the types of feminist texts we had been reading, toward something radical and boundary stretching. We were mentoring the dissertation toward the imagined future we saw for the other's ideas, and always coming back and making kin to their philosophies. Their values, their ideas, and ours were also intermingled.

Relational becomings, staying with the trouble

Haraway (2016) described tentacular tangles of making kin through "the patterning of possible worlds and possible times, material-semiotic worlds, gone, here, and yet to come" (p. 31). In addition to the concept of making kin, Haraway brought us the concept of staying with the trouble. We had both read the book separately before our meeting, and in it, Haraway implored that staying with the trouble "requires learning to be truly present, not as a vanishing pivot between awful or edenic pasts and apocalyptic or salvific futures, but as mortal critters entwined in myriad unfinished configurations of places, times, matters, and meanings" (p. 1). Through our mentored-together relationship, we sorted through advice and feedback, discourses about what it meant to be a graduate student and job-searching and new faculty. In addition to the swarm(s), we invited/curated/traced through feminist theories and texts; there was other noise in the swarm – well-meant advice, handbooks, and job talk requirements – that asked us to become differently. We wondered together, what does it look like to be marketable, to write a good dissertation? Navigating the often contradictory and swirling advice of being-on-the-job market, we did work for each other, sorting through layers of suggestions. Do this; do not do that. Say this; wear this. Become marketable in a humanist sense, "the systemized standard of recognizability – of Sameness – by which all others can be assessed, regulated and allotted to a designated social location" (Braidotti, 2013, p. 26). We were in the swarm moving toward graduation, becoming (hopefully) hired, and also on the periphery working toward posthuman becoming (Braidotti, 2013) in our same spacetimematterings (Barad, 2007). Our relationship outside/inside of the hierarchy allowed us to question some of the unquestioned trajectories and movements that filled the air around us. Mentoring each other toward what appeared to be a stable subject (for the length of a job talk or cover letter), yet holding at the same time that that subject was not that stable at all but always on the move (perhaps even nomadic) (Braidotti, 2006).

> Did you hear from UGA re the Qual position? I am hoping that you did. I got an email right after we stopped talking on Friday. After stewing for the weekend, I decided that it is probably best to be upfront about it. I value our relationship and felt it would be awkward not to say since I knew you applied. I am hoping you got an email on Friday as well and have been stewing too. So . . . here goes. I also understand if you don't want to share either way, and I am glad this is probably the only job we will both be applying for!
> Ohmygoodness, I am glad you reached out! I was thinking about this all weekend as well . . . Yes! I also got an email on Friday, and am happy to hear that you did as well!
> Attached is my research talk . . . I haven't timed myself with it yet, but would love any feedback you have – some of the slides have notes under them, some don't – it's somewhat cobbled together from my proposal presentation and the research talk I did for a class . . .
> BTW please send good vibes, I have a Skype interview for a position that I applied for tonight at 7pm!

These were times when the trading of writing was a formality for connection, and the dissertation texts were pushed aside as we spent our entire meeting storying our lives forward, imagining possible futures and ways to live well in the academy. We were trying to live together well. And we both got jobs and now we find ourselves entering into two very different mentoring situations; two institutions with different philosophies on what tenure track should look like, and we trade stories still, make kin and connect and tether to feel safer on the edge. Affirming each other in small acts of resistance to other well-intentioned advice and mentorship. Swarming, working tensions between recognizable and resistant, experimental and marketable, pliant and rigid.

How do you tell the story of a mentorship? Of a kinship, a meeting, a becoming together? A jump in time and space, sitting across from each other at Susan's dining room table, cooling cups of coffee between us as we write in different parts of this document. What does it do to call this mentorship, we wonder; what is the takeaway we are offering? Perhaps, one of the takeaways we offer is in the ongoingness of our kinship. Mentor-kin-swarm as a relational becoming not bounded by graduate school or positions or a dissertation-to-complete. Rather, as we both now find ourselves navigating the threads and grooves of academic life, we are always already co-constituted by the material, textual, and human interactions of which we are a/part. We make intentional choices to draw lines to each other, to bring each other feminist texts from emails or journals or hallways, to acknowledge the feminist swarm of which we are a/part. Our swarm-mentor-kinship reminds us when we are pulled toward linear trajectories of becoming scholar that there are other futures worth imagining. The swarm and its multidirectionality, its ongoing movement, and interconnectedness allows ventures far afield, excursions and day trips outside the always-on-the-move boundaries. Each excursion stretches the possibility for departure; yet the swarm is always there, and there, and there, as lines of connection expand the feminist territory. Other feminist swarms can break off and hover in careful relation to the particular spaces and times of breeze and scent and temperature and wing. We can be content in the still productivity, the slowness, waiting for scouts to report back on the best site to build another hive. The stillness makes us vulnerable, and it carries potential for new imaginings. As we seek other imaginings, we are affirming difference and multiplicity as allowable within the pack of academia – "tending towards the 'and yet' as a mode of joyful life-affirming 'doing otherwise' in higher education" (Taylor, 2019, pp. 6–7). The exploration of the "and yet" asks for a de-acceleration and a lingering. At this time and others, we (re)turn our mentor texts, read, and reread. We spread and collect the feminist pollen as we journey to other spaces and times, always making kin back/forward in ethical ongoing relation and co-constitution.

Yeah joy – glad it's in both of our dissertations. We have to write out of this co-dissertating in some way . . . or write something together – after the defenses of course!

I love this as an ending and an invitation.

References

Acker, S., & Dillabough, J.-A. (2007). Women "learning to labour" in the "male emporium": Exploring gendered work in teacher education. *Gender & Education, 19*(3), 297–316. https://doi.org/10.1080/09540250701295460

Ahmed, S. (2012). *On being included: Racism and diversity in institutional life*. Duke University Press.

Ahmed, S. (2017). *Living a feminist life*. Duke University Press.

Aiston, S. J., & Jung, J. (2015). Women academics and research productivity: An international comparison. *Gender & Education, 27*(3), 205–220. https://doi.org/10.1080/09540253.2015.1024617

Ali, S., & Coate, K. (2013). Impeccable advice: Supporting women academics through supervision and mentoring. *Gender & Education, 25*(1), 23–36. https://doi.org/10.1080/09540253.2012.742219

Barad, K. (2007). *Meeting the universe halfway: Quantum physics and the entanglement of matter and meaning*. Duke University Press.

Barad, K. (2013). Ma(r)king time: Material entanglements and re-memberings: Cutting together apart. In P. R. Carlile, D. Nicolini, A. Langley, & H. Tsoukas (Eds.), *How matter matters: Objects, artifacts, and materiality in organization studies* (pp. 16–31). Oxford University Press.

Blum, L. D. (2010). The "all-but-the-dissertation" student and the psychology of the doctoral dissertation. *Journal of College Student Psychotherapy, 24*(2), 74–85. https://doi.org/10.1080/87568220903558554

Braidotti, R. (2006). *Transpositions: On nomadic ethics*. Polity Press.

Braidotti, R. (2011). *Nomadic theory: The portable Rosi Braidotti*. Columbia University Press.

Braidotti, R. (2013). *The posthuman*. Polity Press.

Brooks, S. D., Dean, A. S., Franklin-Phipps, A., Mathis, E., Rath, C. L., Raza, N., Smithers, L. E., & Sundstrom, K. (2020). Becoming-academic in the neoliberal academy: A collective biography. *Gender and Education, 32*(3), 281–300. https://doi.org/10.1080/09540253.2017.1332341

Cannon, S. O. (2018). Teasing transcription: Iterations in the liminal space between voice and text. *Qualitative Inquiry, 24*(8), 571–582. https://doi.org/10.1177/1077800417742412

Cannon, S. O. (2019). *Doing science: Data enactments in mathematics education and qualitative research* (Publication No. 77, Doctoral dissertation). ScholarWorks@Georgia State University.

Cannon, S. O. (2020). Making kin with comprehensive exams: Producing scholar in intra-action. *Qualitative Inquiry, 26*(1), 36–50. https://doi.org/10.1177/1077800419869962

Dohm, F., & Cummings, W. (2002). Research mentoring and women in clinical psychology. *Psychology of Women Quarterly, 26*, 163–167.

Flint, M. A. (2019). *Methodological orientations: College student navigations of race and place in higher education* (Publication No. 2241659855, Doctoral dissertation). ProQuest Dissertations & Theses Global.

Flint, M. A. (2020). Fingerprints and pulp: Nomadic ethics in research practice. *Art/Research International: A Transdisciplinary Journal, 5*(1), 1–15. https://doi.org/10.18432/ari29485

Flint, M. A., & Guyotte, K. W. (2019). Pedagogies of the minor gesture: Artful mentorship in college teaching. *Visual Inquiry, 8*(1), 63–75. https://doi.org/info:doi/10.1386/vi.8.1.63_1

Grant, B., & Knowles, S. (2010). Flights of imagination: Academic women be(com)ing writers. *International Journal for Academic Development, 5*(1), 483–495. https://doi.org/10.1080/136014400410060

Grant, C. M., & Simmons, J. C. (2008). Narratives on experiences of African-American women in the academy: Conceptualizing effective mentoring relationships of doctoral student and faculty. *International Journal of Qualitative Studies in Education (QSE), 21*(5), 501–517. https://doi.org/10.1080/09518390802297789

Guttorm, H. E. (2016). Assemblages and swing-arounds: Becoming a dissertation, or putting poststructural theories to work in research writing. *Qualitative Inquiry, 22*(5), 353–364. https://doi.org/10.1177/1077800415615618

Hancock, T., & Fontanella-Nothom, O. (2020). Becoming with/in flux: Pedagogies of Sustainment (POSt). *Qualitative Inquiry, 26*(1), 81–88. https://doi.org/10.1177/1077800419874827

Haraway, D. J. (2016). *Staying with the Trouble: Making Kin in the Chthulucene.* Duke University Press Books.

Haraway, D. J. (2018). Staying with the trouble for multispecies environmental justice. *Dialogues in Human Geography, 8*(1), 102–105. https://doi.org/10.1177/2043820617739208

Hughes, H. E., & Vagle, M. D. (2014). Disrupting the dissertation, phenomenologically speaking: A reflexive dialogue between advisor-advisee. In R. N. Brown, R. Carducci, & C. R. Kuby (Eds.), *Disruptive qualitative inquiry: Possibilities and tensions in educational research* (pp. 243–260). Peter Lang.

Jemisin, N. K. (2018). *How long 'til Black future month?: Stories.* Orbit.

Kelley, M., & Salisbury-Glennon, J. (2016). The role of self-regulation in doctoral students' status of All But Dissertation (ABD). *Innovative Higher Education, 41*(1), 87–100. https://doi.org/10.1007/s10755-015-9336-5

Liao, M., Pegorraro Schull, C., & Liechty, J. M. (2009). Facilitating dissertation completion and success among doctoral students in social work. *Journal of Social Work Education, 45*(3), 481. https://doi.org/10.5175/JSWE.2009.200800091

Locke, L. A., & Boyle, M. (2016). Avoiding the A.B.D. Abyss: A grounded theory study of a dissertation-focused course for doctoral students in an educational leadership program. *The Qualitative Report, 21*(9), 1574–1593. https://nsuworks.nova.edu/tqr/vol21/iss9/2

Magano, M. D. (2011). Narratives on challenges of female Black postgraduate students. *Teaching in Higher Education, 16*(4), 365–376. https://doi.org/10.1080/13562517.2010.546523

Myers, K. D., Cannon, S. O., & Bridges-Rhoads, S. (2017). Math is in the title: (Un)Llarning the subject in qualitative and post qualitative inquiry. *International Review of Qualitative Research, 10*(3), 309–326. https://doi.org/10.1525/irqr.2017.10.3.309

Pope, E. C., & Edwards, K. T. (2016). Curriculum homeplacing as complicated conversation: (Re)narrating the mentoring of Black women doctoral students. *Gender & Education, 28*(6), 769–785. https://doi.org/10.1080/09540253.2016.1221898

Shelton, S. A., Guyotte, K. W., & Flint, M. A. (2019). (Wo)monstrous suturing: Woman doctoral students cutting together/apart. *Reconceptualizing Educational Research Methodology, 10*(2–3), 112–146. https://doi.org/10.7577/rerm.3673

Shelton, S. A., & Melchior, S. (2020). Queer temporalities, space time matterings, and a pedagogy of vulnerability in qualitative Inquiry. *Qualitative Inquiry, 26*(1), 51–59. https://doi.org/10.1177/1077800419868507

Taylor, C. (2019). Unfolding: Co-conspirators, contemplations, complications, and more. In C. Taylor, & A. Bayley (Eds.) *Posthumanism and higher education: Reimagining, pedagogy, practice and research* (pp. 1–27). Palgrave Macmillan.

Van Cleave, J., & Bridges-Rhoads, S. (2013). "As cited in" writing partnerships: The (im)possibility of authorship in postmodern research. *Qualitative Inquiry, 19*(9), 674–685. https://doi.org/10.1177/1077800413500932

VanTuyle, V., & Watkins, S. (2010). Voices of women in the field-obtaining a higher education faculty position: The critical role mentoring plays for females. *Journal of Women in Educational Leadership, 8*, 209–221.

THE MENTOR I NEVER KNEW I NEEDED

Rebekah R. Gordon

Image 1.1 The Mentor I Never Knew I Needed

Source: Rebekah Gordon; Photo credit: Mark A. Kakuba

 DOI: 10.4324/9781003022558-4

2

FLIPPING MENTORING

FEMINIST MATERIALIST PRAXIS AS QUIET ACTIVISM

Carol A. Taylor

Introduction: re-turning, aerating

A number of years ago, a colleague and I developed the Relational-Ethical-Affective-Dialogic (READ) model of mentoring (Nahmad-Williams & Taylor, 2015). This model emerged from a two-year, institutionally supported mentoring project and was based in a shared desire to move beyond traditional, hierarchical, expert/novice approaches to mentoring. The model we developed aimed to capture what we saw then as a potentially transformative approach to mentoring, which shifted from dialogue as a verbal mode of information transmission during dyadic mentoring meetings to the relational, ethical, and affective dimensions of dialogue. This shift, we felt, reconfigured mentoring as embodied academic praxis. The project mattered a lot to both of us. It took place at important points in both of our careers: I was a becoming-senior research leader in my department with a growing interest in helping junior colleagues navigate the stresses and strains of building a career in academic contexts often inimical to feminist collaboration; my colleague was an early career researcher who, while having as many years' experience as a lecturer and teacher as I, had recently obtained her PhD and was looking to shape up a publication plan, publish her first article, and develop ideas for her future research.

In the neoliberal terms that we both critiqued in our paper but found institutionally necessary to adhere to, the project was an undoubted success.

DOI: 10.4324/9781003022558-5

We wrote and published a paper together; my colleague drew up a five-year publication plan; we had regular meetings; we drew up an action plan with goals and a timeline; and we shared expertise in relation to our desire to escape from such neoliberal exigencies. We wanted mentoring to be time to get to know each other and space to explore ideas away from endless and exhausting busy-ness – to have some fun thinking together. Looking back now, I am struck by what our READ model missed: the material dimensions of mentoring are, if not entirely absent, pushed into the background. This seems odd, not just because *now* the mattering of materiality is part of how I think-breathe-live, but also because we were aware of materiality's import. For instance, we referred to our empirical materials as "material artefacts" (Nahmad-Williams & Taylor, 2015, p. 186), and we deliberately chose to meet in a café, not an office, because we felt the matter of the space in which mentoring happened was important. In addition, while our paper, in a somewhat traditional vein, outlined "key themes," from the start we questioned what "data" "is" and troubled traditional notions of data gathering and analysis. All this was part of our collaborative reaching toward a sense of mentoring-writing together as emergent material practice. There are probably many reasons why materiality was not featured as strongly as it might: the push to produce a paper in a particular timescale; the journal word limit, which too often constrains academic articles into "small boxes" which contain "so much and no more" thinking; and the focus on developing our model. Whatever. Materiality, and materiality's entanglements, were (I now feel) insufficiently attended to in the published paper.

A focus on materiality, then, is what propels this chapter, offering a springboard for a re-turn to the project and for a new passage in/through theory-data. This re-turn is not about a simple going back to the project to bring its fugitive materiality to "light." Such an approach would imply that the data has been shelved, stored, and is simply lying there in an inert state waiting for me to bring it to "life" again so that I might honor it with my human gaze, divest it of its analytical insights, and squeeze it dry of meaning. Rather, I propose a re-turning inspired by Karen Barad (2014) who figures it as:

> A multiplicity of processes, such as the kinds earthworms revel in while helping to make compost or . . . turning the soil over and over – ingesting and excreting it, tunnelling through it, burrowing,

all means of aerating the soil, allowing oxygen in, opening it up and
breathing new life into it.

(p. 168)

Re-turning as "aerating" is not a merely reflective reviewing of past events,
but as Benozzo et al. (2019) explain, is a "dynamic and generative [pro-
cess of] invigorating past/present/future connections" (p. 89). This chapter
deploys the re-turn as a strategic practice of aerating, aimed at producing
new thinking about the mentoring project by focusing more specifically on
how its lively materialities flip it into a praxis of quiet feminist activism.

The following section lays out the philosophical terrain of/for flipping
mentoring. The chapter then moves through three relational flips, each of
which aerates different material aspects of the mentoring project. These
flips are: flipping care, flipping muddle, and flipping time. It ends with the
argument that flipping mentoring is about doing mentoring differently so
that it comes to matter as an entangled feminist praxis which has a quiet
but effective power in the increasingly toxic times of the neoliberal academy.

Flipping mentoring

Relational flips

I recently took up Massumi's (2015) ideas on "relational flips" as a means to
consider how to resist normative modes of research (Taylor, 2020), but the
notion of relational flips continues to provoke me: is this idea also useful for
reconceptualizing mentoring? This is another Baradian re-turn – it aerates
the concept again in ways which helps instantiate Deleuze and Guattari's
(1991/ 1994) point that "the object of philosophy is to create concepts that
are always new" (p. 5).

In *The Politics of Affect*, Massumi (2015) offers a rethinking of power as
an affective political ecology of resistance. Power, he argues, is not "external" to us but is a mode of affect that traverses bodies, working relationally
between and amongst bodies to enable things to happen differently than
usual. Power, he suggests, "in-forms" us; it emerges with us through relations
that actualize affective flows in-and-between bodies. Power, rethought by
Massumi, is a form of affective relational agency which carries the potential
for freedom – for escape from what appear to be closed, deterministic systems. Mentoring in neoliberal mode is, I think, one such closed system: it
relies on an expert/novice dyad; it is hierarchical; it presumes a linear notion

of self and career development; it presumes progress to an "inner circle" of knowers; it is outcomes-focused; and, while it does often include self-reflection and reflexivity, it remains largely "monological" and impervious to innovative meanings or constructions (Bokenko & Gantt, 200, pp. 247–248). This monological mentoring model presumes power travels in one direction, from mentor to mentee, and – while it may often be a "successful process" – it is oriented not towards change but rather towards confirming existing normative institutional practices. It is also often geared to ensuring that the mentee's development, albeit central, remains bound within normative neoliberal practices of outcomes, productivity, and self-promotion.

In contrast, Massumi's conceptualization of power as a potentializing freedom which "in-forms" us bodily and produces possibilities for escapes, unpredictabilities, and errancies, seems now a more philosophically fruitful way to consider the affective relational flows and the often micro-material matterings that shaped the mentoring events my colleague and I engaged in. Massumi (2015) refers to these potentializing routes of escape and errancy as "relational flips," which, he suggests, can push us into "totally different places than you'd expect" (p. 17). The shift from traditional ideas of power as repression, constraint or power over, towards a positive sense of power *to*, is also made by other philosophers who have theorized power as productive: Foucault (1980), of course, but also Deleuze and Guattari (1980/ 1987) and, much earlier, Spinoza (1677/ 1996), and feminist philosophers such as Grosz (2005) and Stengers and Despret (2014). What drew me to Massumi's notion of "relational flips" was that it enables a generative way of tuning into what became affectively actualized in/through/by materialities and matterings when our bodies-in-relation collided in the mentoring project. Flipping mentoring in feminist materialist mode is not something you can do alone; it is not part of the machine of neoliberal individualization and self-responsibilization. Flips are relational, and, as Massumi makes clear, the errancies activated by flips emerge in ecologies of practices. They produce affective moves in a politics of change that occur in events, happenings, and unfoldings.

Quiet feminist activism

The "potentializing" freedom that relational flips enable point to the importance of – the mattering of – micro-dissident feminist practices instantiated in the doings and undoings, the thinkings and unthinkings, the learnings and unlearnings, that became possible in the intensive in-between of the mentoring project. Such matterings are often barely perceptible events and happenings that were navigated as difference emerged and materialized in our conversations, doings, and writings. "Event," in this sense, is allied with Deleuze's (1969/1990) concept of the event not as something which disrupts the normal flow and which "stands out" (which is what "event" usually implies) but as a series of ordinary moments of ongoing and often minute transformations in which forces and interactions are imminently actualized. Such processual and "indeterminate events-in-the-making" emphasize flipping mentoring as a "practice . . . that by definition never concludes but results in a living of the theories to which one subscribes" (Kuntz, 2020, n.p.) The theory/ies we were/are "living" was/is an entangled multiplicity oriented to doing mentoring differently. *Then*, we wanted to do mentoring against the grain of neoliberal mentoring via the development of READ. *Now*, I am re-turning the project with feminist materialist philosophy, thereby to explore its feminist impetus as a mode of quiet activism.

Why quiet activism? My colleague and I "got on with" our project. We laughed as we cocooned ourselves in the thrill of having allocated institutional time to develop a practice of feminist mentoring as a minor art of subversion from within. We smiled when we met to talk and think and write. And sometimes, when the moment opened out, we shared a look, a raised eyebrow, a sigh, a glance – a shiver in the present/presence of a happening felt relationally as a shared sensorial-spatial experience that needed no words. We felt no need to "proclaim" the value of what we were doing, or to engage in demonstrations of its value to others. We used our time to be and do together – to create a time-space of air and light to work out together how to do mentoring otherwise. I see this *now* as a form of quiet feminist activism, which refuses both the dominant definition of activism as representative participation in the public sphere, and the binary that places activism in opposition to inaction. Instead, our quiet feminist activism engaged a politics of nomadic companionship oriented to acknowledging the value of material moments (Taylor, 2018) of rupture which released opportunities for creativity and subversion. Such quiet feminist activism nevertheless has material valence and affective force that enabled us to move counter to the neoliberal forces of apathy, negativity, and cynicism. It shored us up to fight the same-different battles that feminists continue to fight in institutions where toxic masculinities can rampage; it was an affirmation of the work that feminists continue to do at micropolitical levels and in assemblages and alliances in opposition to patriarchal stories and histories and violences. Flipping mentoring as quiet activism was (I *now* realize) a mode of what Donna Haraway (2016) calls "ongoingness" – a feminist means of staying with the specific trouble at hand of mentoring in neoliberal institutions. Forging ongoingness is necessary and difficult, and one needs traveling companions!

Where, then, is my traveling companion (my colleague)?

Why am I writing this chapter alone? The project was a long time ago, and I now work at a different institution; our paths have diverged as we have forged different career trajectories and become entangled with different theoretical and practice communities. When writing the chapter draft, I emailed and obtained permission from my colleague to mention her, but I made the decision that the re-turn I wrote (this chapter) would use only my diary entries (which follow), not hers, although she gave me permission to do that. The quote of hers I include in this chapter is from our published article together and is already in the public domain (Nahmad-Williams & Taylor, 2015). Yet, throughout my writing in this chapter, she remains with me as colleague-companion, her presence felt as an affective resonate as I wring new writing from my thinking "re-turn" and attend to the project's (ongoing) matterings. Lindy's email refers to the uncertainties we face in this not-yet-post-COVID times, and she tells me she will "go with the flow." That makes me smile. It sounds so right, so like her.

My colleague's absent-presence is another thread in the mattering of the pleated text that is the mentoring project. It requires attending to the ethics deriving from ongoing response-ability. I figure "response-ability" here as a mode of regard – particularly, of regard which activates a knowing entangled with care, concern, kindness, respect, and admiration. Regard, in this sense, deepens relations. Re-turning mentoring with regard, then, is about looking at how what we did was oriented to efforts to refuse expert/novice hierarchies. Instead of my colleague "learning from" me (as in transmission belt mode), such mentorship instates a sense of learning-together-with, creating space for our emerging, felt sense of the mutuality – of a relationality – that might move in the direction of equalizing esteem. There can be no end to regard and the ethical matterings it entails.

Different differences matter

As mentioned earlier, the re-turns in this chapter aim to give matter its due in mentoring as feminist praxis. In this, it follows a new material feminist orientation grounded in the ethico-onto-epistemological approach outlined by Karen Barad's (2007) agential realism and, later, puts some philosophical concepts from agential realism – in particular, "intra-action," "cut," "entanglement," and "apparatus" – to work to explore the mentoring project from a new angle. Agential realism is a major line of thinking within the recent new materialist turn. Fairchild and Taylor (2019) offer the following brief encapsulation of Barad's approach:

> Agential realism, based on the insights that nothing exists in and of itself, that everything is always already in relation, and that matter and discourse are co-constitutive, has been widely taken up as a paradigm-shifting analytical move that works through the challenges posed by quantum physics to Cartesian epistemology and the humanist ontologies that underpin it. Barad's lexicon – agencies, intra-action, entanglement, the cut, phenomena, apparatus, diffraction – deriving in part from the language of quantum physics, offers social science researchers a new range of conceptual resources for putting agential realism to work to investigate the world in new ways. Central to agential realism is the necessity of developing an ethico- onto- epistemological stance that entangles what humanist approaches have illegitimately disentangled.
>
> (n.p.)

What Dolphijn and van der Tuin (2012) say about new materialism's value as a new and radical philosophical approach can be applied to agential realism too:

> New materialism . . . does not add something to thought (a series of ideas that wasn't there, that was left out by others). It rather traverses and thereby rewrites thinking as a whole, leaving nothing untouched, redirecting every possible idea according to its new sense of orientation.
>
> (p. 13)

Re-turning to the mentoring project with Barad's agential realism gives me the opportunity to think of mentoring as a "material – discursive nondualistic whole in which differential patterns of mattering emerge and come into being" (Fairchild & Taylor, 2019, n.p.). It urges me to notice some things I missed, that got elided and excluded the first time around. This re-turn provokes new data cuts in the mentoring project data's potentially unbounded profundity and provokes "a [new] commitment to understanding which differences matter, how they matter, and for whom" (Barad, 2007, p. 90).

Such Baradian-inflected noticings are important in a number of respects. One, this approach brings to the fore matter's agency. Matter is not a passive object, ground or background awaiting "use" or "interpretation" by human agents in social discourses. I therefore pay attention to the entangled and dynamic nature-culture of the material relations, practices, and ecologies of differential understandings of what matters in mentoring. Two, agential realism's presumption of matter's agency shifts ontological presumptions: agency is not a property of individual human bodies; rather, agency is ontologically distributed across material apparatuses that include humans and nonhuman entities, all of whom/which have differing forms and modes of agency. Agency, considered thus, emerges in events (see the previous section) as an intra-active materialization of/in differing forms and forces. And three, agential realism figures a different epistemological approach to researcher relationships and the research process: knowing, Barad (2007) says, "does not come from standing at a distance and representing but rather from a direct material engagement with the world" (p. 49). Re-turning our mentoring project enables me to take a second (and provokes a third . . . a fourth) look at what occurs when we rethink mentoring via the agential realist view that "bodies in the making are never separate from their apparatuses of bodily production" (Barad, 2007, p. 159). With these philosophical orientations to hand, I re-turn to the mentoring project anew.

Flipping care: response-able kin-ships and risky rubbings

The paper we wrote together out of the mentoring project (Nahmad-William & Taylor, 2015) mentions a particular university café three times, my "office" twice, and offices as institutional places where mentoring normally occurs once. Why do these mentions matter? We wrote, then, that the spaces/

places of mentoring "took us beyond saying 'the mentoring took place here' and towards the perception and experience of mentoring as a relationally embodied spatial event" (p. 190). Okay so far. But, in this re-turn, I want to explore the spacetimemattering (Barad, 2007) of mentoring a little more: to think mentoring as events of coming-to-matter in complex choreographies of vital materialities entailing objects, bodies, affects, and spaces, whose dynamics helped co-constitute bodily materializations and shape the ethical relationalities mobilized in the project.

Being out of office mattered. My colleague found the café "less intimidating" than my office as a place to discuss her writing drafts; I found it a relief to get away from endless emails, distractions, "to-do" lists, and performative busyness. We reveled in the slower time of "time out" and "time away" and in the materialities of the café's murmur, the coffee's aroma, the cake's taste, and how softer chair and table design enabled different – closer and together – bodily orientations. In this spacetimemattering, talk became more intimate, open, and conversational: my colleague and I became secret-sharers whose smiles spoke of conspiracy. The mentoring task – that I would support my colleague to write the first paper from her recently completed PhD – in the event became enfolded with specific material densities and highly charged affective relays, unspoken and tangibly felt. A brief extract from my first diary entry reads:

I caught at the fizz of possibility, felt a bit heady with the responsibility (oh, god, she's relying on ME!!!), and got totally fired by the brand shiny newness of it – for her, for us, together.

And from Lindy's diary:

I lack confidence in my ability to "be" an academic. Perhaps I should just concentrate on writing an article rather than seeing it as something that gives me an identity or something that defines who I am. It is just an article – keep things in perspective.

The piece of writing I sent was basically cut and pasted from my thesis with a few added bits to join it all together. I didn't want to

spend time re-working it in case it was not the right level of content
for a journal. I await Carol's feedback with trepidation.

Barad's (2007) concept of the "agential cut" refers to the analytic practice of separating "something" – an object, practice, person, for example –
out from the ongoing flow of spacetimemattering which, at the same
time, entangles us (as researcher or writer, for example) ontologically with
the phenomena. For that reason, cuts, Barad (2007) explains, "cut-to-
gether-apart" in that they (a) produce "something" to be analyzed, and
(b) hold those producing the cut accountable for the cuts (the "separa-
tions") they make. I (that is, the "me" writing this "now") am responsible
for the cuts made in the production of this chapter. Thus, this re-turn to
the café to ponder the project is not simply a return to a context or environment. Rather, it makes a cut of spacetimemattering to focus on cof-
fee-cake-table-chair-bodies-article draft-talk. These material particulars
matter; they were instrumental in enabling vulnerabilities to be shared.
This cut continues to resonate: its affective density propels it into the
present "now" in an ongoing ethical holding-to-account. This cut chan-
nels Barad's (2007) comment that "bodies do not simply take their place
in the world . . . rather 'environments' and 'bodies' are intra-actively constituted," demonstrating that "intra-action" is about the co-emergence of
things-in-relation, each of which comes to matter differently but each
of which – human and nonhuman – are agentic within and constitutive
of material reality (p. 170). Each intra-action is also at the same time an
ethics-in-relation.

In mentoring as usual, care is usually thought of as a linear, hierarchical caring "for" or "about." The relational ethics of our project enabled
care to figure differently – as an uneasy entanglement, a situated politics
in which our differential positionings vis-à-vis article writing and academic identities had to be (and continue to be) reckoned with. In material
feminism, care is an enactment, something we do in encounters layered
with fears and foibles (Puig de la Bellacasa, 2017). My colleague's "trepidation" about writing, and her anxieties about "being academic," alongside my concerns about being responsible *for* her writing, produced new

intra-active possibilities including enacting care as kindness, as a generosity that swirled between us, assembling a kinship of mutual regard. Haraway (2016) suggests that "it is high time that feminists exercise leadership in imagination, theory, and action" to make new kinships across borders and boundaries of selves and species (p. 102). In a quiet way, our project did this; it helped "stretch the imagination [to] change the story" (p. 103) so that we could flip mentoring-as-usual. Enacting feminist leadership differently – as quiet activism – shifts responsibility *for* to response-ability as a capacity to respond which displaces notions of the "expert," exemplifying that "it matters how kin generate kin" (Haraway, 2016, p. 103).

I was amused, in reading for writing this chapter, to come across a neuropsychological study on peripersonal space (PPS), that is, space immediately surrounding the body. The study reports on an experiment showing that "sharing experiences with others . . . can lead to qualitatively different spatial representations around our bodies" (Maister et al., 2015, p. 457). The authors suggest that their study provides scientific verification that this re-mapping and expanding of PPS "may facilitate behaviours aimed to protect our partner from harm [and] could play an important role in empathic behaviours, protection and altruistic helping" (Maister et al., 2015, p. 460) – something that my colleague and I discovered empirically in rather a different, but no less important, way in our mentoring encounters, and something that new materialist and process philosophy understandings of relational ethics have been suggesting for rather longer.

Mentoring is a risky process of self-exposure, of opening oneself to another's judgment. It is laced with the fear of one's own insufficiency. My colleague wrote about her "apprehension," about wanting my "approval," about how my "positive feedback" would assure her of her credibility as an academic. I worried about what I might do to her "PhD baby." So, how might response-abilities of kinship flip mentoring away from the ontological terror of being discovered to be "not good enough"? Phillips and Taylor (2009) define kindness as "the ability to bear the vulnerability of others" (p. 8), while Magnet et al. (2014) see kindness as a micropolitical practice of/for coalition building. Kindness, in the project, was about trying to make connections to push back (together) against legacies of shame to refuse the diminishments that attend women's work. But kindness as kinship is a risky rubbing: it eases but it also sometimes chafes; it does not erase power or difference, although it can create space for difficult conversations about those things. Such chafings are irruptions – they materialize as affective concern, anxiety or worry; they weigh on us and compel a response (Shaviro, 2008). In the project, chafings came to matter as relational botherments, rather than being contained as "problems" the inexperienced mentee would take to their mentor to help them "solve." In the relational rubs of kin-ship, as Stengers (2005) notes, you must "pay . . . attention as best you can" (p. 188).

Flipping into the muddle: writing relationally

This material feminist re-turn makes me realize how much time my colleague and I spent "in the muddle" – which turned out to be not such a bad place. Haraway draws affinities between the aerating activities of earthworms as they turn and re-turn soil and the practice of "staying with the trouble" which produces assemblages of "situated work and play in the muddle of messy living" (Haraway, 2016, p. 42). The mentoring project was an experiment in such situated muddlings.

One muddle concerned the impetus that underpinned the project: I would help my colleague shape an article from her doctoral thesis.

Slept well till 2.30am then wide awake so I got up and worked with Lindy's draft till after 6am this morning. Great potential story but where are you Lindy? It's written in an objective style and there's no "I" in it? I could hear Lindy at the corners, in little phrases, tones, and turns of speech, but she was mostly muted by the impersonal . . . need to escape this straightjacket. I have a laugh out loud moment at the deadpan tone in which she writes about the nil response to her last-ditch efforts at using the internet to obtain research participants . . . And then, wow, it just happened, serendipity, fate, intervene and it comes together – whoosh, now the research gets going, within the story of the research which has already been going a while . . . And she's off. And so am I. I work on her draft, over it, under it, through it, imagining myself in her research shoes in the moments she passes through.

Donna Haraway (2016) says it matters what stories tell stories. *Then,* I crafted a rather fanciful story to explain my colleague's research journey: I wrote grandiosely in my diary entry, "The Trial: Lindy's Kafkaesque battle with the prison service to do her research!" *Now,* I invite a more modest story, a story about how muddling-together was a generative space for discovering the joys of un-chunking, un-constraining, and unboxing knowledge and for coming-to-know writing as a tentacular practice.

Engaging together with "The PhD" (capitals important!) revealed its form and force as a constraining artefact. Doctoral theses are written documents produced to long-codified institutional norms, a form explicitly designed to address and display standards of research quality, and an artifact for producing oneself as a recognizably knowing-knower able to inhabit academia with the revered title of "Dr." "Extracting" the potential article from the constraints of the PhD and producing it as a "recognizable and ready" journal article took six months of muddling-together, of to-ing and fro-ing, of writing more and writing over and into, of discomfort, disruption, and disassembling of the "known" that had been solidified during the years of doctoral pedagogy. *Now,* I see that the notion of "extracting" doesn't work as an explanation of our processual, writerly meanderings: the article was not a tooth that could be taken out whole. It was a slow finding, a care-full crafting, which put new things in motion and that crisscrossed research, time, and space anew. Rather than an extraction, the emergence of the article was a "re-moving" (Rheinberger, 1994, p. 78) that choreographed new body-writing-space-matter-assemblages in the specific relationalities that came-to-matter in that event of mentoring. Such writing as "re-moving" is a "contingent iterative performativity" (Barad, 2014, pp. 173–174). Hepler et al. (2019) use the lovely phrase "edging with mist" to talk about the process by which writing happens, and of texts which are not yet "but somehow must look as if they are" (p. 145). Undoing the fixed solidity of "The PhD," re-moving it, was exactly that: an "edging with mist" that disassembled to materialize something else anew.

The second generative muddle was writing the paper that eventually became our joint published article. On Thursday, 23 August 2013, I wrote this:

Source: Carol A. Taylor

W e are always in the middle and, as D and G say, "the middle is where things pick up speed." But endings, however fake, do routinely occur in institutions as we discipline our minds and bodies to "tidy that away" (that old module, course, year, person, role, desk) and "move on" to the next thing. So, I had the great pleasure just now of "closing down" my "SPDT Mentoring" folder and opening a new one called "2014 Mentoring paper first draft" in my "Articles" folder. I know it's only 2013 but the aim is to get the article written and published during 2014. Our mentoring project is – a woven interweaving – an up yours to this ending-beginning nonsense. The article will emerge in and amongst our many coffees and ongoing conversations. It's beginning is already past . . . we'll talk about it and something will materialize out of the talking, the reading and the doing. And so it did. Yesterday, as we meandered, wondered and wandered about immersed in "what did we do – what will we do?" questions, a series of words emerged to focus and structure our future writing: Respect, Words, Voice, Process, Diary. I suggested a "theoretical framework" (Buber) to help us think about dialogic mentoring as an ethic of relation. Lindy assuaged my fears that my diary was navel-gazing by suggesting that we each responded to the concept and form of "diary" individually and very differently and that was ok and we write about in the article.

We met fortnightly initially, then when the article began to take shape, less frequently. We wrote sections separately, shared them in advance, then spent the time together mustering, mashing, mushing, pushing, cajoling, summoning, hailing, and inviting words to assemble on the page. Every . . . single . . . word, comma, semicolon, full stop, phrase, sentence, paragraph was written over together, discussed, agreed on, and approved by both, in a collaborative writing process that was a slow, care-full mutual holding to account. We eventually stopped tinkering. The disorderly painstaking work over many months produced sufficient words assembled in a useful order that we were happy enough with to send off to peer review. Doing writing like this – as a form of rhizomatic nomadism traversed by affective resonances in which something emerges from the muddle – felt of a piece with the relational ethics of the project. The false starts, the dead-ends, the round and rounds that eventually became ever-decreasing circles, mattered as much as the "something" which eventually materialized as the article. Thinking-feeling-writing as material practice mattered. We bricolaged something and, as Loch et al. (2017) note, "who wrote what does not matter, it is not important. Words flow with intensity, they leak, and they cry out . . . they create spaces to do things differently . . . They produce joy. They enable joy – joyousness" (p. 72). And joy is central to the possibilities quiet feminist activism might produce for doing mentoring differently.

Flipping time

This chapter contains numerous statements about "then" and "now." These words are there to help clarify events for the reader. As I indicated earlier, they do not emanate from a reflective, or even reflexive, review of past events – a later holding up of a mirror and a looking back along time's arrow to what happened before as if the mentoring project and its materials were "there" simply waiting to be gazed at again. Such a view would be an indefensible straightening that the analytic practice of the Baradian re-turn abjures. Instead, the re-turn I make works with a sense of time-space-mattering as a scrumpled geography (Doel, 1996) of foldings and enfoldings. The mentoring project continues moving and "re-moving," its affective affinities pulse and bind anew, its pleats and creases continue to emanate forces that entangle and activate new intensities and insights. In this third re-turn, I consider the after-lives of the empirical materials produced in the project and the hauntings its diary-entries-as-data continue to provoke.

I glance through the diary entries as I write my chapter, jogging my memory:

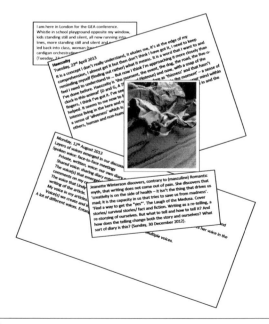

Source: Carol A. Taylor

These entries set off trails, effusions, bursts, flashes, and emanations, entangling themselves forcefully with presents and possible futures:

Post-lockdown, Bath is quiet, tourists remain cautious. Days marked by anger and anguish and worry. A venal government run by posh public school boys for the self-serving upper class. What racist, colonialist nightmare am I trapped in? A dissipated social security system. A de-funded NHS. Protest neutralized by a decade of the grinding poverty of austerity. I am home. How lucky. I have some outside space and a big enough home. I have the resources to "stay safe." In UK, Covid-19 has rampaged through care homes. The government failed to supply adequate PPE across public services. The elderly appear to be the most easily expendable. Brutal. Shocking. True. The highest death toll in Europe. Fearful as schools re-open and workers are chided by the PM to "go back to work" to what we know are unsafe office environments. Our University labs are cautiously re-opening. Colleagues around the world prepare for online or hybrid teaching. Jobs for millions have gone. Academic jobs are at risk. Student numbers unknown. New lockdowns in some cities. Trepidation for a winter Covid-19 spike. In amongst, I feel the need to stave off the dead, cold hand of cynicism and re-member the joys of academic collaboration and creativity. Difficult. But . . . the mentoring project was (is) one such time-space, a matter for joy in which incipient feminist imaginaries made fires that made a little trouble and rainbows that forged a little hope.

I wonder . . . perhaps the mentoring project's flips, its work of relational affiliation, partakes of the quietness of spider work, a tentacular spinning of strong, fine silk. I think now it offers a strategy with which to launch and throw caution to the wind.

I wonder . . .

(3 September 2020).

Lisa Blackman (2019) talks of data's "after-lives" and of how data acquires "agencies that extend and travel beyond the original event or transaction" (p. xxi). The empirical materials of the mentoring project appear as traces, hauntings, and fragments. These flashes and fugitive pieces invite speculations on the project's ongoing deviations and aggregations, on how the traces of its past materialize in the present and bring into play affective modalities of "disadjustment, destabilization, and disjointure" (Doel, 1996, p. 421). Barad (2010) says: "The past is not closed . . . 'past' and 'future' are iteratively reconfigured and enfolded through the world's intra-activity" (p. 261). The mentoring project continues to flip time *now*, and in all the new nows that are on their way, bringing the temporal relation of then/now into indeterminate, recursive coalition.

Concluding somehow now

So what?

The "so what" question is, in positivist research, traditionally considered as the "killer punch question," aimed to ensure the student/novice/ researcher/writer explains the "value" of the research/inquiry/work they have undertaken and spells out clearly its key messages and impacts. Many in the "posts" terrain, myself included, refuse such reductive accounts of the ends and outcomes of their research. Nevertheless, and yet: we (in the "posts") do need to engage in questions of how what we have done/do matters and how our doings-thinkings-writings might encourage or provoke future doings-thinkings-writings. I can only speculate, but here are a few insights that may be gleaned from the above re-turns and flipping mentoring explorations:

- Quiet feminist activism suggests a philosophy-practice of living theory so that we can engage in micro-level but significant dissident practices that work against the grain of our capture by the repressive and negative forces of neoliberal systems and procedures.
- A feminist materialist account of mentoring refuses egoistic individualism in favor of an affective relationality, felt and shared in ongoing processes of ethico-onto-epistemological matterings.
- Mentoring as quiet feminist activism creates a space – one that is risky and affirmative – in which we can express unknowings and explore vulnerabilities and fragilities, so that we might unlearn and learn differently.
- It suggests a need for mentoring leaders in qualitative inquiry to do their leadership differently to enable their students and mentees

not to be caught in, pinned, and traumatized by the deadening lures cast by the urge to mastery.

- It suggests the need for students to be brave, take heart, plunge, flail, cast about, make a splash and make a noise, to engage in experimental and productive disconcertion, to practice curiosity, engage speculation, work with wonder, and not to be disciplined in these doings by the cruelty of the normative gaze which categorizes such behavior as "aberrant."

- Quiet feminist activism can, I think, create a space of/for mentoring as a line of flight for creative collaboration in which the process, what emerges-in-relation, and the support of having a companion in the trouble that (undoubtedly) arises in feminist efforts at staying with the trouble, are what matters.

The chapter has argued that "flipping mentoring" offers a generative and more generous space through which feminist work as quiet activism can be done and have effect in undoing neoliberal modes of mentoring that privilege skills, outputs, and outcomes; that orientate mentoring to self-promotion; and that, through the expert/novice dyad, produce the same. Instead, in attending to the material, affective, sensory, and bodied aspects of human-nonhuman relations, and the ongoing matterings of past/present/future connections, the task of a philosophically oriented mentoring is "not to take responsibility for, to take on the burden of what is, but to release, to set free what lives . . . [to make life] light and active . . . affirming in its full power" (Deleuze, 1962/1983, p. 185). In that sense, flipping mentoring opens to an intensive in-between in which new matterings emerge, and the glimmer of a new politics becomes possible. Regard. Behold. Lo and behold.

References

Barad, K. (2007). *Meeting the universe halfway: Quantum physics and the entanglement of matter and meaning*. Duke University Press. https://doi.org/10.1215/9780822388128

Barad, K. (2010). Quantum entanglements and hauntological relations of inheritance: Dis/continuities, spacetime enfoldings, and justice-to-come. *Derrida Today, 3*(2), 240–268. https://doi.org/10.3366/E1754850010000813

Barad, K. (2014). Diffracting diffraction: Cutting together-apart. *Parallax, 20*(3), 168–187. https://doi.org/10.1080/13534645.2014.927623

Benozzo, A., Carey, N., Cozza, M., Elmenhorst, C., Fairchild, N., Koro-Ljungberg, M., & Taylor, C. A. (2019). Disturbing the AcademicConferenceMachine: Post-qualitative re-turnings. *Gender, Work and Organization, 26*(2), 87–106. https://doi.org/10.1111/gwao.12260

Blackman, L. (2019). *Haunted data: Affect, transmedia, weird science.* Bloomsbury.

Bokenko, R., & Gantt, V. (2000). Dialogic mentoring: Core relationships for organizational learning. *Management Communication Quarterly, 14*(2), 237–270. https://doi.org/10.1177/0893318900142002

Deleuze, G. (1983). *Nietzsche and philosophy.* Columbia University Press. (Original work published 1962).

Deleuze, G. (1990). *The logic of sense.* Columbia University Press. (Original work published 1969).

Deleuze, G., & Guattari, F. (1987). *A thousand plateaus: Capitalism and schizophrenia.* Continuum. (Original work published 1980).

Deleuze, G., & Guattari, F. (1994). *What is philosophy?* Columbia University Press. (Original work published 1991).

Doel, M. A. (1996). A hundred thousand lines of flight: A machinic intro-duction to the nomad thought and scrumpled geography of Gilles Deleuze and Felix Guattari. *Environment and Planning D: Society and Space, 14*(4), 421–439. https://doi.org/10.1068/d140421

Dolphijn, R., & van der Tuin, I. (2012). *New materialism: Interviews & cartographies.* Open Humanities Press.

Fairchild, N., & Taylor, C. (2019). Karen Barad: Game changer. In P. A. Atkinson, A. Cernat, S. Delamont, J. W. Sakshaug, & R. A. Williams (Eds.), *SAGE research methods foundations.* SAGE. http://methods.sagepub.com/foundations/barad-karen

Foucault, M. (1980). *Power/knowledge: Selected interviews and other writings 1972 – 1977.* Harvester Wheatsheaf.

Grosz, E. (2005). *Time travels: Feminism, nature, power.* Duke University Press. https://doi.org/10.2307/j.ctv125jkvq

Haraway, D. J. (2016). *Staying with the trouble: Making kin in the Chthulucene.* Duke University Press.

Hepler, S., Cannon, S., Hartnett, C., & Peitso-Holbrook, T. (2019). Undoing and doing-with: Practices of diffractive reading and writing in higher education. In C. A. Taylor & A. Bayley (Eds.), *Posthumanism and higher education: Reimagining pedagogy, practice and research* (pp. 141–151). Palgrave Macmillan. https://doi.org/10.1007/978-3-030-14672-6

Kuntz, A. (2020). Standing at one's post: Post-qualitative inquiry as ethical enactment (Special issue "Global Perspectives on the Post-Qualitative Turn in Qualitative Inquiry"). *Qualitative Inquiry, 27*, 215–218. https://doi.org/10.1177/1077800420932599

Loch, S., Henderson, L., & Honan, E. (2017). The joy in writing assemblage. In S. Riddle, M. Harmes, & P. A. Danaher (Eds.), *Producing pleasure within the contemporary university* (pp. 65–79). Sense Publishers. https://doi.org/10.1007/978-94-6351-179-7

Maister, L., Cardini, F., Zamariola, G., Serino, A., & Tsakiris, M. (2015). Your place or mine: Shared sensory experiences elicit a remapping of peripersonal space. *Neuropsychologia, 70*, 455–461. https://doi.org/10.1016/j.neuropsychologia.2014.10.027

Magnet, S., Mason, C., & Trevenen, K. (2014). Feminism, pedagogy, and the politics of kindness. *Feminist Teacher, 25*(1), 1–22. https://doi.org/10.5406/femteacher.25.1.0001

Massumi, B. (2015). *Politics of affect.* John Wiley & Sons.

Nahmad-Williams, L., & Taylor, C. A. (2015). Experimenting with dialogic mentoring: A new model. *International Journal of Mentoring and Coaching in Education, 4*(3), 184–199. https://doi.org/10.1108/IJMCE-04-2015-0013

Phillips, A., & Taylor, B. (2009). *On kindness.* Farrar, Straus and Giroux.

Puig de la Bellacasa, M. (2017). *Matters of care: Speculative ethics in more than human worlds.* University of Minnesota Press.

Rheinberger, H.-J. (1994). Experimental systems: Historiality, narration and deconstruction. *Science in Context, 7*(1), 65–81. https://doi.org/10.1017/S0269889700001599

Shaviro, S. (2008). *Self-enjoyment and concern: On Whitehead and Levinas*. http://shaviro.com/Othertexts/Modes.pdf

Spinoza, B. (1996). *Ethics*. Penguin Classics. (Original work published 1677).

Stengers, I. (2005). Whitehead's account of the sixth day. *Configurations, 13*(1), 35–55. https://doi.org/10.1353/con.2007.0012

Stengers, I., & Despret, V. (2014). *Women who make a fuss: The undutiful daughters of Virginia Woolf*. University of Minnesota Press.

Taylor, C. A. (2018). What can bodies do? En/gendering body-space choreographies of stillness, movement and flow in post-16 pedagogic encounters. *International Journal of Educational Research, 88*, 156–165. https://doi.org/10.1016/j.ijer.2018.02.001

Taylor, C. A. (2020). Flipping methodology: Or, errancy in the meanwhile and the need to remove doors. *Qualitative Inquiry, 27*, 235–238. https://doi.org/10.1177/1077800420943513

LOST IN THE BECOMINGS

Carlson H. Coogler

The thing about "being" a mentor is that it's not really something you can "be." "Being" a mentor is really an entanglement: a situated, co-constructed, temporary intra-action (Barad, 2007), a becoming-mentor that also already includes becoming-mentee. One is not mentor, one becomes-mentor/mentee (Deleuze & Guattari, 1980/2018). Always coupled, always transitioning: smooth space becoming striated, and striated smooth. In my own life, I have been/become student-and-mentee/mentor, teacher-and-mentor/mentee, teacher-and-mentee/mentor, administrator-and-mentor/mentee, student-and-mentor/mentee. These roles fold over each other now, the relations still there (residual, sticky) even when no longer codified or institutionalized. Mentorings-and-menteeings, folding over me and through me, pulling together relationships and moments and roles in a fluid, fractal entanglement.

This is part of the beauty of mentoring. If mentorship (*noun*) is transitional and multiple, it is also the loopings of processes: of becomings-mentor, becomings-mentee, becomings-both partially and simultaneously and surprisingly. *Verb*. Roles slipping and sliding according to the event (Manning, 2013). Making space for surprising and unusual becomings.

I experienced one such becoming-mentor/mentee while embroidering. Embroidery is a fascinating and provocative artful method for me, one that continually evokes thoughts about relation and experimentation. Embroidery generally happens on a piece of fabric stretched on a hoop or frame and involves passing a needle with thread back and forth through the fabric. As it is generally imagined, this technique produces a "front" and a "back." What one thinks of as "the embroidery" is what is termed the "front." In comparison, the embroidery artists typically hide the back, even going so far as to seal it from view. It is not imagined as part of the art. It is not part

 DOI: 10.4324/9781003022558-6

of what one shows. This division and covering creates clear boundaries: the art side, the work side; the side to show, the side to hide; the product side, the process side.

I was thinking about this binarization – and the binarization of mentorship into mentor and mentee, knower and learner, guide and guided – as I prepared to stitch. What would it do to think mentor/mentee not as separate but relational and fluid? To make a front/back that is not static but relational, fluid? A front/back where neither side is covered because both matter, both contribute to the art? A front/back that was . . . front-back-front-back-front . . . ? (And, likewise, a mentor who was . . . mentor-mentee-mentor-mentee . . . ?) I experimented. I made a handful of stitches on the "front" side, and then flipped to the "back." I made a handful of stitches on the new "front," then flipped. Repeated. Repeated. Repeated. Flipped. Flipped. Flipped. Over and over, until stable identifications blurred. Was this the front-back-front-back-front-back . . . ? Or the back-front-back-front-back-front . . . ? Front-ing/back-ing. Mentor-ing/mentee-ing.

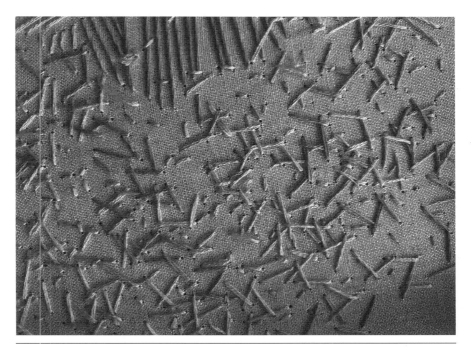

Image 2.1 Lost in the Becomings
Source: Carlson H. Coogler

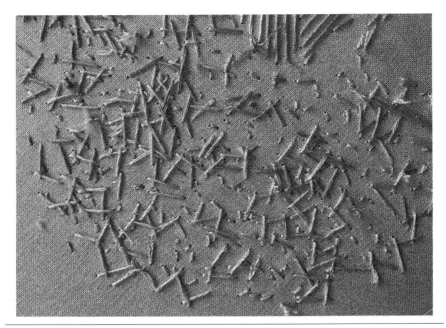

Image 2.2 Lost in the Becomings
Source: Carlson H. Coogler

Here, I show two sections, one on each side, without label. They are both . . . *front-back-front-back-front* . . . Lost in the repetitive flips. Lost in their own becomings-together.

With this experiment, I offer no answers. Just a provocation (Manning, 2013) and a narrative of a time when mentor/mentee, material/idea, embroidery/mentoring, and front/back became-together. So, I end with questions, not answers. What would it do to get lost in such becomings, to think mentorship with this experience? To think mentorship not as mentor and mentor. Nor as mentor or mentee. Not even as mentor-mentee. But as . . . . *mentee–mentor–mentee–mentor* . . . ?

References

Barad, K. (2007). *Meeting the universe halfway: Quantum physics and the entanglement of matter and meaning*. Duke University Press. https://doi.org/10.1215/9780822388128

Deleuze, G., & Guattari, F. (2018). *A thousand plateaus: Capitalism and schizophrenia* (B. Massumi, Trans.). Bloomsbury Publishing. (Original work published 1980)

Manning, E. (2013). *Always more than one: Individuation's dance*. Duke University Press.

3

KINNING AND COMPOSTING

MENTORSHIP/T IN POST-QUALITATIVE RESEARCH

Jonathan M. Coker, Samantha Haraf, María Migueliz Valcarlos, Scottie Basham, Diane Austin, Dionne Davis, Anna Gonzalez, and Jennifer R. Wolgemuth

Introduction

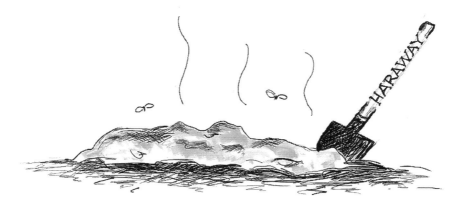

As Haraway's (2015) com-postists (rather than posthumanists), we come together to compost our thoughts intertwined around mentorship in quali-tative inquiry. We engage as com-postists rather than posthumanists so that the "order is reknitted: human beings are with and of the earth" (Haraway, 2016, p. 55). As Haraway notes, prefixes and terms such as posthuman, homo, and anthropos signal a human-oriented view that still places humans apart from everything, whereas compost places us all together in one pile

DOI: 10.4324/9781003022558-7

(Haraway, 2016). As we begin our composting, we consider mentoring as it has been traditionally conceived.

Mentoring is a process through which those new to the field encounter established wisdom, habits of thought, modes of becoming (and more) in qualitative inquiry (Guyotte & Wolgemuth, 2022). As this book demonstrates, mentoring is a philosophical and material practice through which subjects (e.g., mentors and mentees) and their relationships are made manifest. How mentoring is conceived and enacted matters in the ongoing becoming of qualitative inquirers and inquiry, but also in the (re)circulation of mentoring as a powerful and lively concept.

Like all concepts, mentoring is troublesome. Left to its own devices, mentoring risks reproduction of the same, a conservative relationship that maintains oppressive cultures (Leeder & Cushion, 2019), narrow professional standards (Devos, 2010), and strict disciplinary norms and boundaries (Manathunga, 2007). Many of this book's authors argue the point: mentoring is too often depicted and practiced as a dichotomized hierarchical one-way relationship in which knowledge flows from the expert to the novice (e.g., Flint & Cannon, 2022; Guyotte & Marn, 2022). Under this logic of technical breeding, mentoring is a progenitive force of self-reproduction in which the mentee is created in the image of the mentor (Manathunga, 2007) and standard methods and methodological traditions are passed on from one generation to the next. Mentoring's reproductions may pose little trouble for qualitative inquiry conducted in procedural ways (Kuntz, 2016), where there is a separation between subjects from objects, persons from processes, humans from nonhumans, and more. However, mentoring as a process of "making babies" may be especially ill-suited for post-qualitative inquiry, particularly in troubled times where entangled problems like poverty, climate change, and population growth require "making kin" in flat ontological spaces to think "anew across differences of historical position and of kinds of knowledge and expertise" (Haraway, 2016, p. 7). So we turn to composting, because it affirms that we are entangled with these problems. Composting also transforms these entangled problems from waste materials to a rich and nurturing set of materials for thinking.

Inspired to compost rather than reproduce, we wondered what can be said of mentoring, differently conceived as making kin, in qualitative work. How is mentorship understood, enacted, and even discarded in qualitative research in a way that "encourages the creation of the *not yet* instead of the repetition of what *is*" (St. Pierre, 2019, p. 3, emphasis original)? What happens when we conceptualize mentoring in nonhumanist, non-hierarchical, non-traditional ways? How might these conceptualizations foster thinking anew about the possibilities of and for mentoring in qualitative inquiry to speak within the troubles of urgent time?

We (be)came to think with these questions in a new semester-long doctoral seminar on post-qualitative research. We began to understand and appreciate post-qualitative research as a philosophically rich approach to qualitative inquiry with a focus on concepts rather than method. Working post-qualitatively upended the procedural ways of thinking ingrained in us by traditional methods. We were doctoral students and a professor, Jenni, who had learned apart and together in previous qualitative classes, research projects, and reading groups. Although the post-qualitative seminar started in a campus classroom, we restlessly moved through different locations – classroom, bars, restaurants – until we found a place, Jenni's living room, that invited and welcomed us to think with scholars who expanded boundaries for academic scholarship. We consumed food, wine, and post-theorists alongside "mentorship" as a companion concept. The post-orientation of our readings in class challenged traditional notions of qualitative inquiry and paralleled the ways we began to challenge traditional notions of mentoring. It seemed to us that mentoring in qualitative inquiry must be rethought through a post-orientation.

Yet, as we re-thought mentoring, we noticed (and surely failed to notice) how our accounts of mentoring often reproduced what we sought to challenge – dichotomous unidirectional understandings and assumptions – despite our efforts to fertilize possibilities for different conceptualizations of mentoring. We strove to think mentorship through our anti-humanist, posthumanist, and new materialist readings: Haraway, Barad, Manning, Deleuze and Guattari, Foucault, Butler, and more; and we wrestled to resist the old, rusty discourses of mentoring that kept creeping into our thinking. That crafty "mentorship" was guilty of so many humanist crimes: oversimplifying learning, centering humans in learning, creating mentoring subjects and situating them in hierarchical one-way relations – expert to novice, text to reader, teacher to learner, discipline to undisciplined, paid to unpaid.

Our still-forming sensibilities initially turned up their noses at the stench of this hierarchically oriented mentorship, so we called it "mentorshit."

Originally, "mentorshit" denoted our disgust of how traditional mentorship reproduces oppressive hierarchies both among humans (in relation to race, class, gender, ability, and more) but also between humans and nonhumans (Hamilton & Neimanis, 2018). The profanity of this word "mentorshit" expanded our thinking of what could be composted by pushing us to grow beyond a "politeness in the classroom" that often excludes ideas deemed too offensive (Wolgemuth et al., 2020, p. 36). "Mentorshit" helped us to decompose hierarchically oriented mentorship and make kin in the process. In this way, "mentorshit" situated mentoring in Haraway's (2016) terms as making kin in compost. Making kin, for Haraway (and us), is not a linear process but a lateral one in which human and nonhuman entities co-constitute within the humus of hot compost piles. We take the shit – the humus of philosophy and theory – and mulch it with mentoring. As Haraway's com-postists, we pursue mentoring as a "need to make-with – become-with, compose-with – the earth-bound" (p. 102). Just like post-qualitative research orients our research toward creation (Manning, 2016), we seek "mentorshit" that allows us to be experimental, creative, intertwined with subjects and objects, and oriented toward possibility.

This process of composting with "mentorshit" yielded a chapter filled with writing experiments that we started in our post-qualitative seminar. Our writing experiments embrace multi-species (Haraway, 2016), object performative (Barad, 2007), minor (Manning, 2016), and other accounts of mentoring that, composted together, led us to productive uncertainties and disruptions. Our writing experiments with "mentorshit" as making kin and compost include: mentoring-with silences, ghosts, memories, cats, butterflies, mold, and a multitude of other entities in an ecology of qualitative mentoring. By opening mentoring up to its possibilities through kinning and composting, we seek a multiplicity of mentoring that flows with the rhythms and pulses of learning and doing qualitative research.

Although our compost pile has no borders or divisions separating its components, we chose to organize our chapter into three sections around our various writing experiments in relation to scholars that prompted us to expand our thinking around mentorship and post-qualitative research: mentors that helped us decenter humanist approaches (thinking with Haraway, 2016), mentors from the material world (thinking with Barad, 2003), and mentors from the minor gesture (thinking with Manning,

2016). These three sections do not need to be read linearly; instead, our arrangement provides one possible way to examine what emerged from our compost pile.

Mentors that decenter the humanist

Composting mentorship initially overwhelmed us, because we were focused on the human effort needed to compost the hodgepodge of ideas around mentorship. But thinking mentoring through Haraway (2016) leveled us with beings of the earth; and we began to see that, like in any compost pile, the bacteria, worms, insects, ants, beetles, and other critters work in concert to help us to become-with each other. After all, "no being survives or leads 'a life' without that life being sustained by other organisms" (Rautio, 2017, p. 97). Pushing aside the humanist from the center, who had been crowding our thinking and preventing us from noticing our earthly kin, we watched as various nonhuman mentors emerged from our compost pile – cats, butterflies, and more; and we noticed that, "we require each other in unexpected collaborations and combinations, in hot compost piles. We become-with each other or not at all" (Haraway, 2016, p. 4). Here, we share some of the bits and pieces within our compost pile that experiment with wrestling free from the human-centered world and recognizing our previously unrecognized, but ever-present, nonhuman kin as mentors.

Nonhuman animals as mentors

Butterflies as mentors

My initial attraction to Donna Haraway's philosophical perspectives came mostly because of her Camille stories (2016), fantasy literature showcasing a relationship between human and nonhuman, but still both earthly beings. Different, yet with commonality. Sharing a need to survive. A struggle to become. We live. We grow. We die. Hoping to be a better self in the process. But the nonhuman seems to know what it is supposed to do and how it is supposed to live. How is it that we coexist, but nonhumans seem to know so much more than me?

I spent many hours in my backyard sanctuary, planting seedlings, deadheading nectar flowers, caressing and whiffing the rich, pungent soil under my fingernails. Creating a haven for hundreds of monarch butterflies over multiple, seasonal generations. The mariposa. My obsession. My escape. Their wings fragile and ethereal; their lives intense and short-lived. Something I felt I could care for and mentor into being. Do they really need me to guide and support them through their metamorphosis? Is my caring and nurturing for them, or is it for me? If we resist being defined by the construct of species (Rautio, 2017), what are we to one another? And from this wondering that illuminated my human-centered arrogance grew my recognition of mentoring as an entangled process. We are doing/being/becoming together. We help each other in unexpected ways. We are already in the middle of numerous mentor-mentee relationships with our coexisting earthlings – butterflies and more.

A mentoring meeting: Dahlia was a female monarch. After she hatched, she spent several hours preparing for flight, extending her wings. I spent but seconds with her as she took flight. Then there was Chloe. Chloe was a female monarch who went through the same process as Dahlia, except that for some unknown reason, she did not fully dry her wings after emerging. She struggled to move between the flowers, finding nectar to nourish her until she could mate and lay the next generation of monarchs among the milkweed. Or maybe just waiting to be devoured by a large, angry wasp. As her mentor, should I have just crushed the life from her? Would she make anything of herself that mattered?

But I looked at her and said, should it matter that she would not be the biggest, most prolific, or long-lasting of her species during her two-week life cycle? So I spent the next few days moving her from lantana to salvia. Picking her off the ground and holding her up to the sun so she could energetically flap her wings to attempt a take-off. I put her down and she experienced the warmth of the early spring sun on her wings. She kept moving forward toward her goals, making the journey part of her experience. She lived for almost the full two weeks of the life cycle of her mariposa kin.

I read recently about an experiment with flickering light that determined animals perceived time as passing more slowly. And insects, specifically flies, may be seeing us move as if we are in slow motion. Decentering my human perception of time, I see now that her two weeks could have felt like decades to her. Wasting moments may be akin to wasting lifetimes to my butterfly kin. Chloe mentored me to focus, clear my mind, appreciate life, and enjoy the warmth of the sun "on my wings." I learned those days in her company to ask, "What does it matter?" And on delicate wings, my research and writing flits and flutters and lands much more selectively. If we measured in lifetimes how long a research project might take, we might all reconsider what gardens we choose to flap in.

Cats as writing mentors

A moment of stuckness has arrived for the researcher. Frustration builds. The humanist move might be to seek out the human mentors in friends, colleagues, and even articles online. All the while, nonhuman mentors beckon. Finally, the researcher notices them for the first time. They pop out, like seeing the negative space in a drawing making a shape in one of those perception tests. Learning in post-qualitative research involves learning from "mortal critters entwined in myriad unfinished configurations of places, times, matters, meanings" (Haraway, 2016, p. 1). Doing this requires us to decenter ourselves and other humans as knowers and invite animals to be the knowers as others have done with pigeons (Rautio, 2017), piglets (Murphy, 2020), squirrels (Murphy, 2018), and cats (Nordstrom et al., 2020). In this instance, domestic cats offered a path to get unstuck by attending and attuning (Rautio, 2017) the researcher to work-write-think in a shared multispecies rhythm (Nordstrom et al., 2020).

I was home alone watching a Halloween-themed Netflix show, clearly meant for teenagers. The lighting and rain outside added to the ambience. My two cats were cuddled on the couch with me, next to the blanket draped around me. My partner was out of town, so I was home all alone. The television darkened, the music swelled, the rain crashed down, and the thunder struck in concert with the climactic moment on television. I felt something unfamiliar slinking its way up my blanket. I slowly slid the blanket off my legs, and the lightning strike lit up the room to show a lizard staring me right in the face. I screamed. I jumped. I searched. I frantically moved every piece of furniture I could. No lizard. I cannot and will not go to sleep in this apartment knowing a lizard is with me! My cats glared at me, almost as if to criticize my poor hunting methods. I stopped searching and let them lead the hunt. I followed closely behind them. They didn't move so quickly. Instead, they waited, they stared, they gave up and started again. And through this method, they found him. Victorious, I put the lizard outside. Then I sat down at my computer to write. My new cat mentors had taught me all the skills I needed to find something elusive, whether it be a lizard or my muse. So, I waited. I stared. I would give up and start again. And I wrote.

By embracing a different rhythm to the writing, cats led the researcher into a slow ontology. A slow ontology is another way of "disrupting our writing habits" (Ulmer, 2018, p. 734), and can be accessed through nature to gain "natural cycles,

timings, and the rhythms of everyday" to find slowness (Ulmer, 2017, p. 204).
The slow ontology facilitated by the cats eased the pressure off the writer from
working in goal-oriented, efficiency-demanding bursts to working instead in
multispecies rhythms.

Mentors from the material world

As we continue to compost our thoughts on mentorship, reading and thinking
with Barad (2003, 2007) has been a particularly nurturing material. While
turning the soil of our compost, we turned to the material world to see the
force our abiotic co-laborers have been exerting all along. Divisions between
human and nonhuman work are dissolved, and we experience an expansion
of mentors. This inclusion of the material world mentored us to suspend our
notions of individual agency where one force acts independently on the other
through interactions. Instead, Barad's (2003) concept of intra-action points
to the various entanglements we have with spaces, objects, and physical matter of all kinds. The image of a researcher as an independent figure alone in
the room seeking to make things happen is challenged by a reconceptualiza-
tion where the researcher is in a multitude of relationships with all manner
of matter (that matters). Mold in a classroom can guide us to reconsider the
ethical commitments that a researcher has made.

Mold and urine mentor

Barad (2003) compels us to embrace matter again, a thought that continues to gnaw away at my research agenda. It echoes in my head with questions like "How is my research impacting anything that matters?" while
professors and the university press me to publish wherever I can in a race
to legitimize myself on an academic job market that seems indifferent to
me. In those pockets of time where I stare up at the ceiling, stuck from
tiredness or existential dread about the possible futility of the time and
energy I've invested, the question of what matters circles in my head, pum-
mels my mind. Encountering post-philosophy while earning a PhD has me
feeling like I am learning to ride a bike that is slowly disappearing behind
me as I ride. Newly mastered concepts dissipate under searing critiques of
representational thinking. Barad's mattering mentors me back to consider
what drove me to this place, and I am reminded of all the schools in which
I worked.

My classrooms were often windowless cinder block rooms with fluorescent lights. The mold problem in the building ate through posters and books I owned, but our official inspection report told us there was no mold problem. Matter didn't matter. Instead, it was the words on the paper and the signature. Representationalism mattered more than the matter. But we indeed had a mold problem. The mold spread through the school. Teachers and students alike contracted pneumonia at an alarming rate. Someone has to do something! I am someone. I can do something . . . but not as a teacher. I need more power than that, don't I?

Yet, when I arrive at graduate school, I am pulled to write about identity and agency and other individualizing concepts. Data that has already been gathered is highly valued. I experience bureaucratic complications of collecting data from schools. I experience obstacle after obstacle blocking me from researching the place that brought me here. Instead, I am fed a healthy diet of abstractions, theoretical frameworks, and other representations. Barad continues to push me back. I consider all the urine that teachers hold hour after hour with no relief. I think about the acid churning in their stomach from wolfing down food too quickly. I see the stains on their clothes that come from fatigue and not having the energy to do yet another load of laundry. All of this matter indeed matters. I am so removed from it now, but the memory of the musky smell of mold and the weight of urine in bladders keeps nagging me, urging me to check up on the people I've left behind in the classroom. I call old teacher friends. They ask me what I am working on, as if I am someone who can make a difference. I look at piles of paper and obscure readings and then feel feeble. I have to do better than this.

Others have found ways of cutting into this Gordian knot by examining how the architecture of a school produces poor psychological health (Lenz Taguchi & Palmer, 2013), how objects in a school can shape sexuality (Allen, 2015), and even how classroom materials can shape subjective experiences of time (Hohti, 2016). I too must find a way to cut into this material world of the school to find that more "livable school" (Lenz Taguchi & Palmer, 2013). So, I stop fighting that mold and urine from our old school building, as it continues to entangle me with my old teacher friends. Instead, I let it lead me out of this maze and back to helping in a way that actually affects the physical matter in those teachers' lives.

Mentors from the minor gesture

Re-cycling our thinking through Manning (2016) breathes fresh air into our compost pile of mentorship. Like any compost pile, we see there are materials that are less composted and those that are more composted. The dichotomy between the two reminds us of Manning's minor gesture, and we consider the uncommon mentors we may have ignored along the way in our composting process. Our attention is drawn to these minor mentors. We focus on silence over voice, forgetting over memory, and ghosts over the living, embodied. The minor mentors expose the ways in which we chain ourselves to traditional methods rather than seek the unexplored.

Silence as mentor

Thinking with Manning (2016), we sought out mentors not recognized by traditional methods of qualitative inquiry. Manning writes that "method works as the safeguard against the ineffable: if something cannot be categorized, it cannot be made to account for itself and is cast aside as irrelevant" (p. 32). Manning describes the minor gesture as one such thing unrecognized by method. She writes, "The minor thus gets cast aside, overlooked, or forgotten in the interplay of major chords" (p. 1). Engaging with the minor gesture as our mentor, we then open ourselves up to the "cast aside" and we find silence.

My sudden loss of hearing due to lupus drew me into a relationship with silence that I never expected. Over the course of six months, the silence in my life increased as I experienced a 50% decrease in hearing, then 70%, then 80%. I was in the middle of writing a dissertation on student voice, so the irony of losing my hearing was not lost on me. I began to explore and ponder this new silence in my life, initially to resist it but then to find a way to resist voice and make room for silence.

Silence makes me uncomfortable. I feel an internal power struggle making it difficult to even hear silence. I am constantly exposed to sound even in my thinking, breathing, and quiet moments. There is never complete silence. I am always hearing the sound of my own thoughts, memories, and experiences speaking to me. I think about what I choose to say and not say, who I say it to, and where and when. Manning (2016) talks of major and minor inquiries. In my head, the major gesture is what I say, what society, education, family, cultural norms, have enacted upon me, what is acceptable, what is appropriate. The minor gestures are the voices in my head, cussing, screaming, rebelling, challenging, dissecting, questioning, against normative thinking.

Often in traditional qualitative mentoring, "Good methodologists are taught to focus on and analyze what participants talk about, what they tell us, what they describe, what they recount" (Mazzei, 2007, p. xi). In other words, we are trained and mentored to place voice on a pedestal. Can I find a way to resist this mentoring? Mazzei and Jackson's (2019) critique of voice is a scholarly ally of resistance for me. They argue that voice "has become seen almost as a mirror of the soul, the essence of the self" (p. 67). And my voice does not mirror my "soul." It is a fragment of a mirror, shattered, like an image in a carnival fun house, a distorted vision that someone else created – a fragment that allows one to see what they wish to see, obscuring what they do not want to see, and trapping hidden figures inside the glass.

Nordstrom (2015) critiques how recording devices have been hailed as an apparatus capable of capturing reality. Through this, I see how the recording device plays along with the idea that it can capture someone's soul through what one says, showing it as a reflection of the "soul" of the human- ist subject. We are seduced by this shiny prize into completely ignoring what one doesn't say, or what is being "said" through silence – which may be equally valuable in data collection. So, I resist and find Mazzei (2007) who proposes silence not as "a lack, an absence, or negation but rather as an important and even vital aspect of the fabric of discourse" (p. xii). This is why some researchers use transcription markings to note pauses or other voiceless aspects of an interview, and include these elements in their analy- sis (e.g., Kuby & Vaughn, 2015).

I continue to pursue silence, because it has the power to mentor us to embrace the possibilities of the unsaid, the quiet, and the unheard. Silence creates space for voices to be heard that might lead to new understandings of resistance and agency (Manning, 2016). As I continue to explore this new land of silence, I find a kinship in Erling Kagge, a famous explorer who was the first person to reach the South Pole alone. Amidst the silence of an Antarctic land, he wrote "to speak is precisely what silence should do. It should speak, and you should talk with it" (2017, p. 11). My hearing loss places me deeper and deeper into this conversation with silence to hear new things.

Ghosts and memories

With others who expand boundaries for academic thinking, research, and data (e.g., Koro-Ljungberg & MacLure, 2013; Manning, 2016; St. Pierre, 1997) and who conjure conversations between the living and the dead as co-mingling kin (e.g., Malone, 2019; Nordstrom, 2013), we consider the ways that memories and ghosts might function as mentors. Invited by Manning (2016), we peer at the fringes to see what or whom can be found by turning to the imperceptible. Mobilizing ghosts and memories as mentors requires us to release them from a prison of an unchanging past and to work with them in the present to create new futures. Engaging with these mentors suspends us from temporality and opens paths to things we once knew and forgot as well as things others once knew and perhaps still know.

St. Pierre (2014) urges us to read post-qualitative authors when she writes, "One must read and wrestle with texts" (p. 10). We follow this advice to commune with these thinkers as mentors for our thinking in the same way we would with ghosts. We are all constantly being mentored by ghosts. We meet them in books and articles, pleading with us to seek out new evidence for their theories. We hear them, we read them, take notes on them in class. We imagine them talking by asking, "What would [Manning, Haraway, etc.] say about this?" Some of us even imagine whole conversations or debates with these ghosts. Our data and how we interpret this data are haunted. It's not just researchers, either. Methodology is often described like a recipe book. And many Southern cookbooks have the instruction, "Season until the ancestors whisper enough." Many of us have probably eaten food seasoned according to ghosts, and many of us have consumed data analysis influenced by ghosts.

SAM: I don't believe in ghosts. (*Laughs.*)
JONATHAN: (*Laughs.*) I don't think it matters.
SAM: No?
JONATHAN: Whether we believe in them or not, what matters are the ways ghosts as a concept affect people's thinking.
SAM: That's interesting, because I'm thinking about how many grad students we know that talk about scholars [living, or not] as a type of "mentor" in their mind. Even though

they have never met, that person is still influencing their thoughts.

JONATHAN: Right. And what is the difference between that and thinking of ghosts as mentors?

SAM: So (*pause*), a person's conception of ghosts – or maybe their memories of people or past experiences with people – influence them to think or be a certain way, maybe make them feel cared for?

JONATHAN: Yes, I think so, because doesn't everything matter when we think about how we think?

SAM: Should we get out the Ouija board? (*Both laugh.*) No, but I wonder how we commune with these ghosts . . .

JONATHAN: And can our thoughts ever really be attributed to us alone?

SAM: Right. The humanistic arrogance that every thought we think is us and us alone.

JONATHAN: That reminds me of a friend of mine. He was visiting the grave of his mother, on the anniversary of her death. He feels that is when the distance between them was the shortest. He said, "Not sure if it's 'cause I die a little today. Or if it's 'cause she lives a little more." Maybe we have to die a little, or our thoughts have to die a little, so these ghost mentors can live a little more.

SAM: Okay. So maybe we start by letting "our" thoughts die down and letting someone else talk. Do you think a ghost has to be somebody who is dead?

JONATHAN: It doesn't have to be. Who we once were, but no longer are is a type of ghost we leave behind.

SAM: Okay, then. Haraway, anything to say?

HARAWAY: "*I t matters what matters we use to think other matters with; it matters what stories we tell to tell other stories with; it matters what knots knot knots, what thoughts think thoughts, what descriptions describe descriptions, what ties tie ties. It matters what stories make worlds, what worlds make stories*" (Haraway, 2016, p. 12).

SAM: Then our thoughts are never only ours. They are a compost pile, filled with the "old" and the "new," ideas from the "living" and the "dead," with blurred boundaries and impossible distinctions.

JONATHAN: That is so interesting, because it makes me think that our memories are constantly engaged in composting every image, every taste, every sound, every text, every class into one big pile. And we create from that memory compost, we plant new ideas on pages.

SAM: Yes . . . It also makes me think about how we have a tendency to assume that ideas and humans are inseparable. What is an idea apart from a human?

JONATHAN: Maybe an idea can be a ghost.

SAM: Hmm. You know, sometimes I'm writing and I know an idea or concept is influencing my thinking, but I don't remember "who" originated that idea (or maybe, who first articulated it? Maybe it was already thought before that particular person wrote it?).

JONATHAN: Either way, the ghost, the memory, the mentor become with our thinking . . . *In thinking, in writing, in research, in everything* –

SAM: I feel like writing traditional qualitative work has gotten harder and harder. Because I keep thinking these postqual thoughts. They (the ghosts?) write themselves into my work, even when I resist them.

JONATHAN: I know what you mean. It's like I set this intention of swimming in this one direction. But this undertow of thought changes my direction slightly. And so when I arrive at my destination, everyone wants to hear about the intentionality of where I set out. And a part of me feels like I'm faking everything I say to make it align with this surface level direction I claimed to know about ahead of time.

SAM: Why can't we acknowledge that undertow as a part of the process?

JONATHAN: For me, I feel this obligation to be intentional, directing our consciousness towards something. But in reality, I feel as if we are being pulled by the past, the present, your memories, and my memories.

SAM: Yeah, it's like these ghosts and memories exist in a space where temporality is thwarted, suspending the notion of the linearity of time . . .

We continue, considering the ways memories function without our hailing. There are times we try to remember – reflecting, re-reading, reviewing notes – to guide our decision-making; and other times, when memories impose themselves onto us, unasked for, maybe triggered by a word, smell, a sound, an image. A re-reading of transcripts. A return to the field. And suddenly, we are thinking something different, prompted with a new idea, feeling a new emotion. Here, maybe memories function as a (traditional) mentor by helping us progress, grow, prompting the thinking of new thoughts or the consideration of something not yet considered. Thinking mentorship through memories in a com-postist sense allows us to consider how our research becomes with us and our (nonlinear) memories, and everything else with which we intra-act. We (researchers, participants, subjects) tell and retell stories that are (mis)remembered; these accompany us, become data for our research, become-with us.

Conclusion

We arrived at our semester-long seminar class in the hierarchies and dichotomies cultivated by traditional mentoring. Composting allowed us to break down traditional mentoring and arrive at something new. Engaging with post-qualitative research and writing experiments connected us with kin who helped us reject traditional accounts of mentoring in lieu of the compost pile we call mentorshit. We began the composting process by asking ourselves what can be said of mentoring in qualitative research differently conceived. How must mentorship be rethought? Thinking with Haraway, Barad, Manning, and others, we composted to see what St. Pierre's (2019) "not yets" of mentoring in qualitative research might emerge. Adopting multi-species (Haraway, 2016), object performativity (Barad, 2007), and minor gesture (Manning, 2016) accounts of mentoring, we discovered an abundance of post-qualitative mentors that allowed us to create mentorshit.

Earth seems more alive and full with our post-qualitative mentors. Like any compost, this pile of mentorshit and writing experiments represents a fraction of what we discussed in that post-qualitative seminar, even as its memories intertwine with our thinking now and our work on this chapter. Yet composting has introduced us to new mentors. Butterfly gurus of time showed us to be more selective in what research we spend time on, and cats tutored us on how to work in multispecies rhythms to overcome writing blocks. School building mold and teacher urine compelled us to live up to researcher ethics. Silence coached us to analyze our data for the unspoken. Ghosts and memories helped us to acknowledge hidden influences on our thinking and nonlinear ways we work, think, and research.

Mentoring in post-qualitative research "requires making oddkin" (Haraway, 2016, p. 4). As a class we sought out a location and companions that allowed us to think and become-with researchers that both troubled conventional accounts of mentoring and that "stretch the imagination to change the story" (Haraway, 2016, p. 103). Our writing experiments reveal an assemblage less interested in reproducing and more invested in ways of diversifying mentorship through the concept of mentorshit. This mentorshit created through composting has opened us to new ways of thinking rather than drawing final conclusions, providing us with more to be done and more to be thought. We continue to ask ourselves questions even as we finish this. How else might we flatten the mentor-mentee binary? How might the nonhuman mentor matter? How else might "we" remove the "we" from this thinking/writing/doing? What impact do nonhumans have in the process of data gathering and the production of knowledge? These questions, among others, we leave open for ourselves and for others. We remain in the "thick present" (Haraway, 2016, p. 1), continuing to compost and inviting new kin to join our pile.

Postscript
On Post-Myself and [un]learning

We are no longer what we thought we were: We were never only "we."

We have been aided, inspired, and multiplied[4].

Existence is not an

individual

---

4   Deleuze and Guattari (1980/1988, p. 3).

affair;[5]

We are shifty,

becoming–with,[6]

emerging from *intra-actions*[7]

with our innumerable mentors,

with phenomena and matter, humans, and non:

peers, cats, butterflies, spaces, memories, texts, silences, ghosts . . .

What do I function with (now)?

Who (what) mentors us,

becomes with us?

with our research?

It is not

(cannot)

be just "mine."

I am f r a g m e n t e d –

bits and pieces of me, becoming

(with)in other spaces and (with)in other thoughts

and with(in) other machines and intra-acting in many places,

people and spaces and matter (un)making each other,

inseparable, and these (un)mentors

have made it clear that,

---

5 Barad (2007, p. ix).
6 Haraway (2016, pp. 3–4).
7 Barad (2003, p. 815).

usually, I have

thought

too

small

too close,

too safely,

too agreeably,

adhering to structures of PhDness[8]

equating my identity and my learning with [normative forms of]

researchwriting, assembling particular compilations of

lines and spaces to write about particular things,

using concepts and categories

as structures that actually

close off meaning[9]

How might I

write

~~myself~~

into my work

to reveal the

material/discursive/

social/cultural/natural/physical/

8   Myers et al. (2017, p. 313).
9   Derrida (1967/1973).

biological/economic/geopolitical/ / / /

forces that matter? That mentor? That become-with me and my research?

I am not fully available to myself, I am entangled with

everything else, I can be always only partial,

(and what of my "participants"!)?

What is censored?

What is summoned?

and, what is

forgotten?

"I" am always

in the midst of the event[10]

I cannot be located within a stable letter-word, "I"[11]

We are making kin, making compost, (un)mentoring each other.

References

Allen, L. (2015). The power of things! A "new" ontology of sexuality at school. *Sexualities, 18*(8), 941–958. https://doi.org/10.1057/978-1-349-95300-4_3.

Barad, K. (2003). Posthumanist performativity: Toward an understanding of how matter comes to matter. *Signs: Journal of Women in Culture and Society, 28*(3), 801–831. https://doi.org/10.1086/345321.

Barad, K. (2007). *Meeting the universe halfway: Quantum physics and the entanglement of matter and meaning.* Duke University Press. https://doi.org/10.1215/9780822388128.

Butler, J. (2004). *Undoing gender.* Psychology Press. https://doi.org/10.4324/9780203499627.

Deleuze, G., & Guattari, F. (1988). *A thousand plateaus: Capitalism and schizophrenia.* Bloomsbury Publishing. (Original work published 1980).

Derrida, J. (1973). *Speech and phenomena, and other essays on Husserl's theory of signs* (D. B. Allison, Trans.). Northwestern University Press. (Original work published 1967).

Devos, A. (2010). New teachers, mentoring and the discursive formation of professional identity. *Teaching and Teacher Education, 26*(5), 1219–1223. https://doi.org/10.1016/j.tate.2010.03.001.

---

10 Manning (2016, p. 37).
11 Butler (2004).

Flint, M., & Cannon, S. (2022). Becoming feminist swarm: Inquiring mentorship methodologically together. In K. W. Guyotte & J. R. Wolgemuth (Eds.), *Philosophical mentoring
in qualitative research: Collaborating and inquiring together*. Routledge.

Guyotte, K. W., & Marn, T. (2022). Ruinous mentorship. In K. W. Guyotte & J. R. Wolgemuth (Eds.), *Philosophical mentoring in qualitative research: Collaborating and inquiring together*. Routledge.

Guyotte, K. W., & Wolgemuth, J. R. (2022). Introduction: Why methodological mentoring
in qualitative research? In K. W. Guyotte & J. R. Wolgemuth (Eds.), *Philosophical mentoring in qualitative research: Collaborating and inquiring together*. Routledge.

Hamilton, J. M., & Neimanis, A. (2018). Composting feminisms and environmental
humanities. *Environmental Humanities, 10*(2), 501–527. https://doi.org/10.1215/
22011919-7156859

Haraway, D. J. (2015). Anthropocene, Capitalocene, Plantationocene, Chthulucene:
Making kin. *Environmental Humanities, 6*(1), 159–165. https://doi.org/10.1215/
22011919-3615934

Haraway, D. J. (2016). *Staying with the trouble: Making kin in the Chthulucene*. Duke University Press. https://doi.org/10.1215/9780822373780

Hohti, R. (2016). Now – and now – and now: Time, space and the material entanglements of
the classroom. *Children & Society, 30*(3), 180–191. https://doi.org/10.1111/chso.12135

Kagge, E. (2017). *Silence: In the age of noise*. Penguin.

Koro-Ljungberg, M., & MacLure, M. (2013). Provocations, re-un-visions, death, and other
possibilities of "data." *Cultural Studies? Critical Methodologies, 13*(4), 219–222. https://
doi.org/10.1177/1532708613487861

Kuby, C. R., & Vaughn, M. (2015). Young children's identities becoming: Exploring agency
in the creation of multimodal literacies. *Journal of Early Childhood Literacy, 15*(4), 433–
472. https://doi.org/10.1177/1468798414566703

Kuntz, A. M. (2016). *The responsible methodologist: Inquiry, truth-telling, and social justice*.
Routledge. https://doi.org/10.4324/9781315417332

Leeder, T., & Cushion, C. (2020). The reproduction of "coaching culture": A Bourdieusian
analysis of a formalised coach mentoring programme. *Sports Coaching Review, 9*, 273–
295. https://doi.org/10.1080/21640629.2019.1657681

Lenz Taguchi, H., & Palmer, A. (2013). A more "livable" school? A diffractive analysis of the
performative enactments of girls' ill-/well-being with (in) school environments. *Gender
and Education, 25*(6), 671–687.

Malone, K. (2019). Co-mingling kin: Exploring histories of uneasy human-animal relations
as sites for ecological posthumanist pedagogies. In T. Lloro-Bidart & V. Banschbach
(Eds.), *Animals in environmental education* (pp. 95–115). Palgrave Macmillan. https://
doi.org/10.1007/978-3-319-98479-7_6

Manathunga, C. (2007). Supervision as mentoring: The role of power and boundary
crossing. *Studies in Continuing Education, 29*(2), 207–221. https://doi.org/10.1080/
01580370701424650

Manning, E. (2016). *The minor gesture*. Duke University Press. https://doi.org/10.1215/
9780822374411

Mazzei, L. A. (2007). *Inhabited silence in qualitative research: Putting poststructural theory to
work*. Peter Lang.

Mazzei, L. A., & Jackson, A. Y. (2019). Voice in the agentic assemblage. In N. K. Denzin & M.
D. Giardina (Eds.), *Qualitative inquiry at a crossroads: Political, performative, and methodological reflections* [eBook ed]. Routledge. https://doi.org/10.4324/9780429056796

Murphy, A. M. (2018). (Re)considering squirrel – from object of rescue to multispecies kin.
Journal of Childhood Studies, 43(1), 60–67. https://doi.org/10.18357/jcs.v43i1.18265

Murphy, A. M. (2020). No happy endings: Practicing care in troubled times. *Journal of Childhood Studies, 45*(2), 7–13. https://doi.org/10.18357/jcs452202019735

Myers, K. D., Cannon, S. O., & Bridges-Rhoads, S. (2017). Math is in the title: (Un)learning the subject in qualitative and post qualitative inquiry. *International Review of Qualitative Research, 10*(3), 309–326. https://doi.org/10.1525/irqr.2017.10.3.309

Nordstrom, S. N. (2013). A conversation about spectral data. *Cultural Studies ↔ Critical Methodologies, 13*(4), 316–341. https://doi.org/10.1177/1532708613487879

Nordstrom, S. N. (2015). Not so innocent anymore: Making recording devices matter in qualitative inquiry. *Qualitative Inquiry, 21*(4), 388–401. https://doi.org/10.1177/1077800414563804

Nordstrom, S. N., Nordstrom, A., & Nordstrom, C. (2020). Guilty of loving you: A multispecies narrative. *Qualitative Inquiry, 26*(10), 1233–1240. https://doi.org/10.1177/1077800418784321

Rautio, P. (2017). Thinking about life and species lines with Pietari and Otto (and garlic breath). *Trace: Finnish Journal for Human-Animal Studies, 3*, 94–102.

St. Pierre, E. A. (1997). Methodology in the fold and the irruption of transgressive data. *International Journal of Qualitative Studies in Education, 10*(2), 175–189. https://doi.org/10.1080/095183997237278

St. Pierre, E. A. (2019). Post qualitative inquiry in an ontology of immanence. *Qualitative Inquiry, 25*(1), 3–16. https://doi.org/10.1177/1077800418772634

St. Pierre, E. S. (2014). A brief and personal history of post qualitative research: Toward "post inquiry." *Journal of Curriculum Theorizing, 30*(2), 2–19. https://journal.jctonline.org/index.php/jct/article/view/521/stpierre.pdf

Ulmer, J. B. (2017). Writing slow ontology. *Qualitative Inquiry, 23*(3), 201–211. https://doi.org/10.1177/1077800416643994

Ulmer, J. B. (2018). Composing techniques: Choreographing a postqualitative writing practice. *Qualitative Inquiry, 24*(9), 728–736. https://doi.org/10.1177/1077800417732091

Wolgemuth, J. R., Koro-Ljungberg, M., & Barko, T. (2020). (In defense of) pedagogies of obscenity. *Power and Education, 12*(1), 23–38. https://doi.org/10.1177/1757743819850853

FOR KWG

Aubrey Uresti

after a particularly long day at the computer – responding to a slew of emails, preparing lessons and videos for bichronous classes, holding online office hours, and attending virtual meetings – rather customary tasks for the newly minted, freshly hired academic during global pandemic

already settled into the last Zoom of the day – distance learning with my first-year graduate students – I smile as I hear that familiar discourse marker, the one that seems to make its way into every virtual space I find myself in this year

mindful of the words uttered – *for me* – and intensely aware of the cadence, tone, emphasis, and body language conveyed, practically visualizing the table defining the transcription conventions

an auditory astuteness and appreciation for the graceful nuances on the conversational floor – an impromptu series of *chaînés* en relevé just as captivating as a perfectly choreographed Pas de Deux – that began as a natural propensity and developed through time, tradition, and training

nevertheless, the ear is rivaled by the eye – with a keen ability to spot the seemingly mundane and jot it among the many musings and sketches found in a researcher's notebook

 DOI: 10.4324/9781003022558-8

and always remembering that telling the stories of those who confide their words, memories, and day-to-day lives requires the utmost integrity, persistence, and humility, of course

thousands of pages deep into hand coded transcripts, innumerable hours spent listening to recordings, and countless months enduring the solitary endeavor of writing the dissertation later, doubt (paranoia?) can foment in even the most well-trained qualitative researcher

ailing self-confidence bolstered through consultation and in recognizing that the legacy of those

all day sessions spent atop bonnie hills, learning in community – hot tea to cut the morning chill; sitting on the floor amid towers of treatises embodying those who soldiered on before; a critical discussion of theory (tachycardia to the qualitative heart); taming the rebellious and unraveling the mysterious in the data; thickening the fictive kinship bonds over a meal – lives within

noting that the true education is more than simply understanding some utilitarian aspect of the trade, that the context is just as important as the content, and that as others rush to the find . . . *for me*, this is the gift of being mentored by understated greatness – one that is given

genuinely and wholeheartedly, where the mentee's curiosity is matched by an equally robust level from the mentor – recognizing the mighty imprint of this enduring, reciprocal treasure

a fleeting reverie that gives pause ((.)) and reminds me to cultivate aloha

Author Note: In the Solomon Islands, the Kwara'ae use the term *fa'amanata'anga* to convey, among other things, teaching. As central to much of my mentor's research, I found it to be a fitting base for this acrostic poem.

4

RUINOUS MENTORSHIP

Travis M. Marn and Kelly W. Guyotte

< < < < > > > >

"It was painful, losing parts of what I thought then were my identity. I had to wonder if they meant anything anymore." Travis said during our Zoom call. He adjusted his computer screen and continued, "I don't want to turn this into a confessional." We laughed together and paused a moment. "No," Kelly finally offered. "I guess my experiences with ruin have perhaps been slightly less life-changing. Or, perhaps, differently life-changing . . ." Her voice trailed off as she considered their identities in relation: white–biracial–woman–man–black–mother–partner. "I've never experienced those particular losses. Those refusals. Will I in the future?"

< < < < > > > >

Together, we come to this writing because ruin has flowed through us both as we navigated our doctoral programs, and now our journeys as professors during the Anthropocene – a time of human-produced ruin on a planetary scale (Ulmer, 2017). We come together as qualitative inquirers who have refused certain humanist practices of conventionality that attempted to situate us in certainty, stability, and tradition – ideals that no longer seem sensible in difficult times, in our "time of violence

 DOI: 10.4324/9781003022558-9

of multiple scales and intensities . . . of ever-increasing environmental catastrophes" (Kuntz, 2015, p. 13). Tradition, certainty, and stability are not, and perhaps never were, sufficient to meet these challenges. Instead, we seek to bring ruin to these concepts, to refuse the humanist subject in our research practices and in our lives – to embrace the possibilities of ruin.

We employ ruin to embrace a more open vision of mentoring relationships in this epoch, to resist replicating the humanist subject in those mentoring practices, and to commit to "making kin" (Haraway, 2016) with the nonhuman – to making kin with "earthly ones, those now submerged and squashed in the tunnels, caves, remnants, edges, and crevices of damaged waters, airs, and lands" (Haraway, 2016, p. 71). Ruin, both in our research practices and in our physical world, enables us to think differently, to produce differently, and to engage in mentorship with a revised ethical vision of responsibility. We engage with mentoring relations in this way to avoid/refuse/resist replication of those familiar humanist relations that reproduce to destroy – that offer ruin as finality rather than productivity. We come together in this chapter to explore how ruin has enacted, and continues to enact, on our bodies as mentors/mentees.

In this chapter, we explore the concept of ruin as a noun (e.g. the "ruins" of inquiry, the "ruins" of Earth), as a verb (e.g. to ruin oneself, to ruin others), as a process (e.g. ruination – the oncoming of ruin), as an admission of loss (e.g. my identity was ruined), as a source of pain (e.g. my ruined academic life), as a desire (e.g. I want others to feel ruined, too), and points between (e.g. what was I before I was ruined?). Much like the ruins of inquiry, ruin can neither be directly "seen" nor "wielded" like a weapon. We can sense ruin in the feelings of "loss," in the moments of transformation with our mentees and with ourselves – in the margins of our "human" "selves." Here we follow where ruin takes us. In this chapter, we "allowed" ruin to seep into our writing and enable our past selves, long gone, to speak, as we also let our dialogue intercede where it felt right – where ruin seemed to almost bubble over, into and through conversation and email. Throughout, we consider ourselves ruined/in ruins/as ruins because of our mentoring relations, and we interrogate what such ruin means, what it makes possible, where it nudges us now and in the future. Let's begin with the concept of ruin.

Ruin/s

"I've been thinking about the process of ruin in mentorship," Travis began, "and I am wondering if we need to theorize the mechanism of action of ruin – the what and how of ruin." "That's a good point," Kelly said, then paused before continuing. "The thing about ruin is that you don't know it until it is already there. Even if we think of ruin as always already there, there has to be something that unmoors you, a shift below your feet that sets you off balance . . . do we ever find balance again?" What balance would mean on a ruined planet in a ruined time was unknown to us.

< < < < > > > >

The notion of ruin/s has been previously described by McCoy (1996), Lather (1997), and St. Pierre and Pillow (2000) in relation to inquiry. According to St. Pierre and Pillow, the ruins are the broken and incomplete promises, the failures of humanist thinking that also create ruinous spaces in education and methodology from which we must work. Ruins, in this way, are material reminders of things that once were, the lingering effects of structures that are no longer stable/standing, and conceptual disruptions of "reality, truth, rationality, and the subject" (p. 10). Drawing from St. Pierre and Pillow (2000), MacLure (2011) has since asked, "Where are the ruins?" (p. 997) as she pondered how we have been working in the ruins of humanism in qualitative inquiry. Asserting that even as post-structuralism has cultivated methodological creativity, it "has often failed to make a *difference* to the mundane practices of research and the kind of knowledge that it produces"; thus, she theorizes other ways to further the ruin, or at least make "some structural damage" (p. 998; original emphasis) to conventional inquiry practices. MacLure proposed a "ruin of representation" (p. 999) in order to further disrupt the ongoing crisis of representation, encouraging inquirers to consider how they can make their research texts stutter (Deleuze & Guattari, 1980/1987), emphasizing productions of affect, relationality, and wonder. In her article, we realize that MacLure is not trying to necessarily *locate* ruin as her question indicates. Instead, she is interested in ruinous qualitative inquiry and how we might come to ruin as a concept that opens inquiry *differently*.

Through the preceding examples, those who "work" in the ruins of inquiry are those who use the "posts" and post-qualitative inquiry as a means of disrupting conventional humanist practices (St. Pierre, 2014) and engage with inquiry during and with the Anthropocene epoch (e.g., Somerville, 2016). Before moving forward, we feel it important to expand on humanism and convention as they always move in relation to ruin, especially as they become important ideas that inform our forthcoming discussions on mentoring relations. In her scholarship on posthumanism, Braidotti (2013) discusses humanism as an Enlightenment-era "doctrine that combines the biological, discursive and moral expansion of human capabilities into an idea of teleologically ordained, rational progress" (p. 13). Indeed, humanism relies on a unitary, rational, and exteriorly located subject that breeds individualism. With a focus on rational thinking and seeking universal truth, one begins to see how humanist ideals have guided much of the inquiry work that precedes us, with randomized control trials and positivism becoming the norm (Grossman & Mackenzie, 2005) and the gold standard (Cartwright, 2007) to which other ways of thinking/doing are/were held. It is practices and conventions such as these that scholars such as St. Pierre and MacLure (among many others) push against, effectively troubling the very notion of a convention that merely perpetuates problematic commitments to research. Such commitments can never "work" in/with the philosophical "posts." The ruins, then, are where those doing "post" humanist inquiry in our discussion of mentoring practices, find themselves: within the "crumbling edifice of Enlightenment values" (MacLure, 2011, p. 997), the structural remnants of something that, as we will soon discuss, gave us a sense of false stability – a false sense that traditional mentoring practices were or could be enough.

Humanism does not speak to the ontological, epistemological, or ethical positions of qualitative inquirers, like us, seeking to work the ruins of Earth, mentoring relations, and the crumbling of Enlightenment notions of who and what is worthy of being human – worthy of agency and of being "counted." Thus, humanism unravels/dissolves/falls to pieces, leaving us among ruins that continuously crumble before our very eyes, a productively unstable ground from which we work/think/inquire/mentor. We at times feel lost, numbed by uncertainty or doubt. The possibilities, though, swirl before us, and we feel compelled by the decaying edifices around us to act and inquire regardless. Where, then, do we go from here? How do we work the ruins of inquiry amidst the ruins of Earth? How does one mentor when one questions their existence as human? As we continue to work through this chapter, we draw inspiration from these conceptions of the ruins, and the push against convention. In fact, we are directly inspired by St. Pierre (2014) when she discussed her early encounters with Deleuze, saying, "For me and others like me, [humanist qualitative] methodology was ruined from the start, though we didn't quite know it at the time" (p. 3). We wonder what it means to be ruined by philosophy, and the ways in which we are always already ruined – ruin becoming not a state we reach, but a process that is always at work. More specifically, we wonder how mentoring as a philosophical practice can be ruinous, and how ruin might be not a negative condition, but one that is affirmative and replete with possibilities and liberation. Before further exploring these questions, we turn to literature on mentoring that creates a necessary, albeit ever unstable, foundation for our conceptualization of mentoring in/as ruin. First, ruin bubbles over, and Kelly maps a genealogy of ruin in an email to Travis.

Genealogy of ruin: Kelly

Friday, January 10th

For me, I became aware of ruin through a series of blips – some noticeable and some imperceptible at the time, and perhaps even now. The moments that unsettled me, disrupted my thinking/doing, were the tiny threads that ensnared me in webs of encounters with philosophy, with myriad other bodies. Enfolding then with now, the moments that became disruptive were the ones shrouded with philosophical discomfort in which my constructivist ways of being were no longer holding with the ways I was becoming in/with the world. The stability of a social world full of interaction suddenly seemed less plausible when intra-actions seemed to abound amidst material discursivity. The idea of stable subjects and stable meanings became a blip when subjects became nomadic and meanings started to slip and slide. The paradigm in which I worked through my dissertation, though meaningful at the time, was not helping me answer the inquiries that pervaded my new thinking, a new ethico-ontoepistemology. Like Chinua Achebe (1994) wrote, things fell apart.

Those tiny threads were at first easy to ignore/brush away; however, like walking into a spider web, even the smallest strands of silk become perceptible when they are plentiful enough. For a long time, I blamed

post-structuralism, then critical theory, then feminist theories, for caus-
ing me ruin, ensnaring me in a web I couldn't, and, in retrospect, maybe
didn't want to, escape. As I think then-now, and as I think the very
theories that inspired me, I no longer see ruin in the concepts and phi-
losophies themselves; I see ruin as something that was always already at
work. Perhaps I, too, was ruined from the start.

– Kelly

Mentoring

"I think, ruinous mentorship happens at conferences, in the hall-
way conversations – in bars and non-academic spaces," Travis says,
then pauses. "I mean, it's that random dive bar at one a.m., and you
find yourself in a tipsy conversation about Simondon or Deleuze,
about identity and loss . . ." We exchange knowing chuckles. "Yes,"
Kelly adds, "the most memorable ruinous moments rarely happen in
conventional spaces, right?" Heads nod. We go on to discuss ruinous
mentorship as happening in-between, in the informal spaces that are
often outside of our offices, or, at the very least, beyond the traditional
configuration of chair-desk-chair. We both wonder if encounters in
such a physical space could cultivate ruinous methodological mentor-
ing, like the experiences we both had as doctoral students, and, now,
have as faculty. Could it proliferate under the conditions of conven-
tionality, under such a guise? "Many of my ruinous mentoring expe-
riences have been sparked during campus walks with mentors, or
over tea with students . . ."

Kelly recounts.
Bodies connecting,
congealing, catalyzing . . .

< < < < > > > >

In our engagements with literature on mentoring, we have been struck
by the ways in which mentoring is presented as a practice easily transfer-
able from one context to the next, with myriad guidelines and "how-to"
resources situated both within and across varied disciplines. The assumptions underlying much of this literature posit that there are specific "best"
practices one might enact in order to cultivate a "successful" mentoring
relationship (e.g., Nick et al., 2012). Further, we have noticed the ways in

which mentoring literature perpetuates hierarchies in which the mentor is positioned as the knowledgeable one, directly influencing the learning and growth of the mentee. As an illustration of this, the APA's Center on Mentoring explains, "A mentor is an individual with expertise who can help develop the career of a mentee" (APA Presidential Task Force, 2006). Through this perspective, mentoring responsibilities require that the mentor take on tasks such as assisting, guiding, preparing, educating, socializing, and supporting, just to name a few (e.g. Rackham Graduate School, 2020). Mentoring, then, is often understood as a one-way practice, flowing from experienced/knowledgeable to unexperienced/unknowledgeable. Thus, the literature sets up clear demarcations between who is actively mentoring and who is being mentored, with clear notions of what responsibilities both mentor and mentee must assume.

The clarity with which much of the literature discusses mentoring relations is muddied when it encounters our uncertain and nebulous conception of mentoring as/with/in ruin. Neither static conceptions of mentoring nor hierarchical power relations are commensurable with our philosophical positioning that takes seriously post-structural conceptions of the subject. To begin, we trouble conceptions of the mentor as superior, knowledgeable, certain, and fixed at a place of success in one's career, and the mentee as innocent, ignorant, uncertain, and "developing." Whereas mentoring as an *a priori* practice that can be captured in a manual might work well for more static humanist conceptions of the subject, we understand ruinous mentoring as only thinkable in/with specific relations as we see subjects as "embodied and embedded" (Braidotti, 2013, p. 22). Additionally, even as we acknowledge that there are practices we have encountered that seem to work well in myriad situations, we would never go as far as to call them "best" practices. "Best," according to whom? What values underlie one's conception of "best"? How do we reconcile the concept of "best" when it creates a category of "not best" or "worst"? "Best" practices simply reinstate conventionality, creating a norm or standard which one must achieve; it is antithetical to the project/practice of ruinous mentorship. Ruinous mentoring, then, resists "best" as an overall goal, perhaps instead striving toward what is desirable in that particular moment/encounter. In other words, "best" practices are not a holistic model for ruin beyond humanism. Rather, we seek ruinous mentoring practices that are fluid, relational, and emergent.

Similar to the ways in which mentor/mentee always shift, slip, and become in relation to one another and to "making kin" (Haraway, 2016) with those traditionally excluded from such relations, we feel it important to make clear that we are not situating ruin in binary opposition to humanism, nor are we creating such binaries between ruinous mentorship and conventional mentorship. Instead, we endeavor to ruin mentorship and allow it to open and proliferate, bringing forth ruinous mentorship as but one of many possibilities in relation to conventional mentorship practices. Thus, conventional mentorship necessarily exists along with ruinous mentorship, much like molar and molecular lines, distinct yet inseparable (Deleuze & Guattari, 1980/1987). Ruinous mentorship might align with lines of molecularity – supple, decentered, deterritorialized – and conventional mentorship aligning with the molar – rigid, hierarchical, territorialized. Through this perspective, we view humanist and ruinous mentorship through their inseparability, their complementarity – that one cannot exist without the other. These relations entail mutual necessity (Barad, 2011) where no discussion of ruin is possible without the discussion of humanist mentoring practices. Ruin is not the abandonment of the human, of the conventional forms of mentorship. Rather, ruin is an asymptotic ideal that drives mentoring relations in productively uncertain ways.

As ruin is a shifting and always oncoming process, we cannot imagine that it has a terminus, some location at which one is finally "ruined." Such a notion runs counter to the idea of ruin as productive uncertainty. As the end of ruin is uncertain, so too is the path of ruin – ruin is intra-actively reconfigured by each material-discursive phenomenon. Ruin is then part of the process of individuation and is responsive to the same material-discursive relations that give rise to particular subjects. With this conception in mind, ruin is not a location at which one can proclaim, "I've reached ruin!" Rather, we often do not realize that we are amidst the ruins until we are "in the middle, between things, interbeing, *intermezzo*" (Deleuze & Guattari, 1980/1987, p. 25). As Deleuze and Guattari (1980/1987) remind us, "Where are you going? Where are you coming from? What are you heading for? These are totally useless questions" (p. 25). Ruin and ruinous mentorship are also not concerned with these questions, nor with tracing a path that took us from here to there. Nor are they about locating the place or time in which we became ruined. Ruinous mentorship is, instead, apparent when we find ourselves swept along the currents and unwilling to grasp the raft of humanism that we realize only offers us false reprieve. The ruins are always already among us; however, we can feel their presence only when we release ourselves to their call.

Hailing the mentee

"What was I before I was 'ruined'? What did I think I was? I am not sure I know any more than I am sure what it means to be 'ruined' now." Travis continued, "Still, I wouldn't want to go back. Would you?"

Kelly thought for a moment and flatly replied, "No."

< < < < > > > >

Mentor and mentee identities, or subjects, are not given or self-evident; rather, both are performative, a series of stylized actions, social positioning, and becomings. These subjects instantiate through hailing (Butler, 1997), a "call" into being. That is, one is "named," addressed, and compelled to respond – a social "demand" that one embodies the subject desired for the target of the call. Though all hailings are incomplete and otherwise imperfect, an individual, upon receiving the call, can recognize themselves in that call and respond by performing, however imperfect, the hailed subject. Each call is the product of a particular material-discursive configuration that renders some performances possible and thwarts others. For the traditional mentor and mentee subjects, these configurations are partly comprised of familiar and popular notions of what a mentor (the "knowledgeable" party) and mentee (the "ignorant" one) are "supposed" to be – an individual is called to embody these discourses – discourses that are inextricably tied to the materiality (e.g., university campus, professor offices, classrooms), imbricated with them. Much like other identities (e.g., female performativity), individuals are expected to perform these preset identities and position others, and themselves, accordingly.

It is instructive to consider a professor seeking graduate student advisees. They hail, or call, students to fill the mentee role instantiated through the power structure endemic to doctoral studies. Only those students capable and willing to fill that role of "mentee" could hear themselves in the call and be able to respond accordingly. As such, only those "willing" to embody and perform this traditional form of mentorship would be capable of responding to that call. It is not that "perfect" subject performances are sought by the hailing, but rather, hailing acts to "tune" (Pickering, 1995) performances, to gesture towards humanist mentoring discourses known to both parties.

In most traditional forms of the mentor-mentee relationship, a hailed mentee is typically an oppressed subject, one for whom ignorance, innocence, and potential (but not emergence) are necessarily always already inscribed. A traditional mentor, as previously discussed, must be the superior subject to the mentee, as the call of an inferior mentor would most likely not be returned. The traditional mentor, holding the humanist "knowledge" of oppression and passing it to those whom these oppressed, creates and recreates the humanist subject. The mentee is colonized by the mentor – put to work by existing humanist structures. These power structures render its subjects likely unable to resist notions of imperialism, capitalism, neoliberalism, and other oppressive forces endemic to the Anthropocene, to the ruin of the physical world. Those who attempt to refuse these forces in academia are exposed to "marginalization, exploitation, and job loss" (Grande, 2018, p. 59). In this way, hailing traditional mentor-mentee relationships in academic spaces can serve to possibly reproduce oppression rather than empower mentees to resist them.

Despite these common figurations of the mentee, this subjectification may be desirable as even injurious subjects are occupied in order to provide a recognizable social identity (Butler, 1997). That these mentor-mentee subjects are continually made and remade in ongoing becomings creates the possibility that they evolve from or are transformed by ruinous relations. Subjects are continually rearticulated, and though injurious in their initial instantiations, they may not remain so. As Butler noted of taking on an injurious subject, it is only paradoxically by "occupying – being occupied by – that injurious term can I resist and oppose it" (Butler, 1997, p. 105). Conventional mentorship may hail and then colonize mentees, but in that attempt, resignification, reterritorialization (Deleuze & Guattari, 1980/1987), may be possible. The hailing, the call, may also similarly change – the conventional call becoming a ruinous one. What the ruinous call may hail could then exist outside of the humanist power structure that hierarchically positions mentor and mentee.

Rather than continually reinscribing the power dynamics that created the possibility of the Anthropocene, the flattened power relations of a ruinous call may allow *sympoiesis*, the making of kin (Haraway, 2016). It could begin to blur the lines between mentor and mentee, to flip them, unravel them, leave the terms themselves obsolete – ruined. The ruinous call and the traditional mentoring call could exist together, heard and acted together to enable possibilities yet to come. It is also possible that what the ruinous call hails may not be unified subjects but something different; perhaps this is the source of the vampires, cyborgs, nomads, tricksters, hybrids, and angels that are "lurking, strolling, or dancing in the ruins of research" (MacLure, 2011, p. 997). The unknown and uncertainty that accompanies the call has the potential to enact on ruinous subjects in painful ways, as we will turn to discuss. Before this discussion, we pause to consider Travis's genealogy of ruin.

Genealogy of ruin: Travis

Tuesday, January 14th

Rather than over time or incrementally, ruin came, initially, all at once to me – I felt its effects without any real knowledge or understanding of philosophy or methodology. I was then a second-year doctoral student with very traditional post-positivist and interpretivist orientations. My conceptions and assumptions about "research" collapsed in that moment of ruin, but ruin quickly seeped into other aspects of my becoming; quickly, almost immediately, that uncertainty spread to my past and my conceptions of who "I" "was." It took months to come out of that process I would now call ruination. During that time, as you say, things fell apart, but not in a chaotic, indiscriminate way. Rather, the pieces formed new conceptions, ones that were no longer in line with my previous becomings. It was like looking into a broken mirror – still me, but with the smaller pieces rearranged so much more easily.

– Travis

Ruin as nostalgia, nostalgia as ruin

"Ruin enacts on bodies in such powerful ways," Kelly offers. "It is simultaneously liberating and terrifying. It's almost like you mourn for the ways things were, even as you don't want to return to them." "It is

terrifying," Travis adds. "It feels like a loss even as it opens so many possibilities. I mourned the loss of my racial identity, but it no longer felt like mine. I'm not sure what else I could have done. In no longer feeling 'biracial,' for a time, I didn't know what to tell others or how to conduct research on 'biracialism'

– an identity that no longer felt 'real' to me."

< < < < > > > >

When we hail the call, we come to see ruinous mentorship creating discomfort for the once-humanist, perhaps even causing pain. We do not see pain or discomfort as something to be avoided – discomfort can render possible new insights in inquiry and instantiate transformation (e.g., Wolgemuth & Donohue, 2006). Pain is likewise a source of possibilities as it configures and sensitizes researchers to realities not previously accessible (e.g., Marn, 2018). Here, we will work through the pain that often accompanies ruinous mentoring, thinking with the notion of nostalgia not as an end point, but as a necessary yet temporary state that moves us toward more affirmative and optimistic possibilities in our mentoring relations.

To be sure, the appearance of stability and certainty afforded by humanism has the tendency to lull us into a false sense of security – a false sense that neither inquiry nor the Earth are "ruined," that humanist practices still "work." We argue that those ideals also seep into conventional mentorship practices. As conventional forms of inquiry are bounded in procedures and notions of representation, so too is humanist mentorship mired in false notions of stability. It is perhaps the attempts of conventional inquiry and mentorship practices to limit or avoid pain and discomfort that demonstrate the need for ruin. What happens amidst the shift away from convention in both qualitative inquiry and in mentorship? What happens when we turn away from security and toward the potentiality in ruin? What happens when pain is productive? What happens when risk, excess, and uncertainty are embraced rather than spurned?

In the ruins of mentorship, there can certainly be mourning for the loss of what-was, a type of nostalgia that looms when we feel ourselves becoming differently. The turn away from convention when we hail the call of ruinous mentorship opens us to new vulnerabilities and often painful realizations. The stability we once felt beneath us – *terra firma*

– cracks, jolts, and pulsates, throwing our bodies into a state of uncertainty; we find our footing is unsteady, and we struggle because we are losing what we have always known. We have only ever felt the stability associated with convention so we often ache to (re)turn to what-was. This is what happens in ruinous mentorship. The ache, the pain: we become nostalgic as we think back to those times before we knew we were ruined, and sometimes we yearn for the very thing we tried to escape. This is the double-bind – we know conventional mentorship no longer philosophically holds; however, we are fearful and/or anxious of what lies ahead. Thus, we loop between what-was and what-is-not-yet, exhausted in our becoming-ruined states.

Braidotti (2019) explains, "Fear, anxiety and nostalgia are clear examples of the negative emotions involved in the project of detaching ourselves from familiar and cherished forms of identity" (p. 255). When some think of ruin, it is often perceived in the negative: as broken, wrecked, scarred. Something about a body is marred in some unalterable way, it will never return to the same. However, we see the inability to return to the same as positive in that it keeps us moving toward the not-yet as a space always to be envisioned and never to be fulfilled. Such uncertainty, as we have discussed, can be painful. Considering the roots of nostalgia, meaning "homecoming" and "pain," we wonder what nostalgia might mean if we focus on the inability to return to the same, if home is not a stable place because it never *was*. In other words, what if we are actually ruined from the start? What if the pain is not because we have left a "home" but because we come to discover that "home" is, in fact, the failed promises of humanism? What, then, do we do with nostalgia?

If we (re)think our final question, we are actually more interested in what nostalgia is doing *with us*. Braidotti (2019) creates a way forward with nostalgia in which it becomes a necessary part of the process of ruin; however, it is a pause, not a stopping point. In ruinous mentorship, nostalgia is the space where we come to terms with our loss, the space where we grieve. Approaching nostalgia through a nonlinear conception of time, we realize that nostalgia, as a concept, cannot hold. If we are not operating linearly, what-was never truly was. Thus, we reimagine the pain of nostalgia as simply a layover to an undetermined destination. It becomes a critical space where we mourn the loss of the conventional – be it methodology or mentoring – yet, we do not stop there. Braidotti (2006) explains this well:

The qualitative leap through pain, across the mournful landscapes of nostalgic yearning, is the gesture of active creation of affirmative ways of belonging. It is a fundamental reconfiguration of our way of being in the world, which acknowledges the pain of loss, but moves further. Ultimately, it is a practice of freedom. That is the defining moment for the process of becoming-ethical: the move across and beyond pain, loss and negative passions. Taking suffering into account is the starting point, the real aim of the process, however, is the quest for ways of overcoming the stultifying effects of passivity, brought about by pain.

(p. 84)

In ruinous mentorship, we understand nostalgia as a space of temporary stuckness, and we must resist the lure of the Siren who coaxes us to remain there indefinitely. As alluring as it can be, nostalgia is a false end. The promises of nostalgia are fulfilled in stasis, in looking backward, in the comforts of conventionality. Instead, like Braidotti's (2011) nomadic subject, we see value in acknowledging our pain, but we choose to allow our homes to move with us. Thus, we never truly "come home." The pain, then, is not associated with homecoming but with traversing the "mournful landscapes of nostalgic yearning" (Braidotti, 2011, p. 84). Pain, and nostalgia, are necessarily part of the process but not our final destination in ruinous mentoring. Instead, we also always look forward. We become.

Ruinous mentorship brings an ethical obligation for mentor/mentee to collectively navigate the painful landscapes of nostalgia while seeking the more affirmative quests that always lie ahead, those never fully realized. Rather than be stuck in the nostalgic landscape, we remind one another that nostalgia is a false reminder of what-never-was, and we keep moving. We, mentor/mentee, move together. As Braidotti (2013) asserts, "Affirmation, not nostalgia, is the road to pursue" (p. 10). In the next section, we consider more fully how ethics is at play in ruinous mentorship, and the implications for our mentoring relations as mentors/mentees.

The ethics of ruin

"The world is ending," Travis proclaims in our virtual chat. "We cannot just do what we have always done . . ." Throughout our conversations, we often joke that Travis brings the pessimism to Kelly's optimism. With the

current backdrop of our writing being the worldwide novel coronavirus pandemic, we find ourselves sheltered-in-place and uncertain about an end to what everyone keeps telling us is our "new normal." Kelly's optimism has faded a bit over these weeks, and, to her chagrin, Travis's pessimism seems to poignantly reflect the stark reality that bears down upon them both/all. Together, we pause in our conversation to mull over ruinous mentorship, its effects and affects. We have realized that we work well together in this writing because, in order for ruin to be productive, it requires optimism, even as it acknowledges the challenges that always present themselves, shifting and uncertain as they may be. "An optimism through our pain and exhaustion," Kelly explains. "We need others to get through this – whatever this is. An ethical commitment to relations, to connection, that's what ruinous mentorship requires. Connection is what we all need right now."

< < < < > > > >

With nostalgia as an ever-present undercurrent in ruinous mentorship, we are acutely aware of the ways in which pain, uncertainty, and discomfort can flow between and through mentors and mentees. The very notion of ruin as often meaning something negative and undesirable requires those who practice ruinous mentorship to attend to what ruin does to all bodies involved. If we hail ruin, if we invite ruin in, we (as mentors and/or mentees) must pause to consider the implications of what ruin makes possible and how it enacts in ruinous mentorship. Thus, the mentor/mentee relation must acknowledge and attend to (even as it can never reconcile) the ethics of ruin, which is what we now turn to discuss.

In thinking ruinous mentorship, we are particularly inspired by the ways in which Braidotti (2019) and Haraway (2016) discuss ethics as a practice that zigzags through various bodies, as well as their optimistic and affirmative stances on ethics in this precarious time. Importantly, they believe recognizing the ever-present yet shifting challenges that we face in our current moment, finding ways to move with them rather than negating their existence. Similarly, our ethical stance is one that takes seriously the challenges of moving within a ruinous landscape: understanding the presence of pain disguised as nostalgia, and the lure of humanism that lurks like a shadow, ready to welcome us "back" to a place we do not remember nor do we wish to (re)turn.

In turning toward ruinous mentorship, we become intricately entangled in a practice that defies convention, linearity, and traditional hierarchies. Thus, without distinct and firm boundaries that both separate and define the roles of mentor or mentee, we find ourselves (re)creating mentor/mentee in each moment, in every encounter. Ethically, we realize this asks a lot of both mentors and mentees as they map rather than trace (Deleuze & Guattari, 1980/1987) their relations from within their respective subject positions – positions that shift and evolve. There is no "how-to" manual for this type of mentorship, as it is immanently and philosophically catalyzed, boundless and always becoming. How, then, do we come to terms with a mentorship practice that is never the same? How can we, ethically, invite others into ruinous mentorship knowing the work that lies ahead?

To begin answering these questions, we feel it is important to (re)turn to conventional humanist mentorship. Even as we have focused on the challenges within ruinous mentorship, conventional mentorship also carries the possibilities of pain, discomfort, and uncertainty – difficulties still abound. While ruinous mentorship is certainly a disruptive practice in relation to conventional mentorship, it is important to remember that disruption, similar to ruin, is not an inherently negative enactment. It is also important to remember that this disruption is only thinkable in situating conventional mentorship as the norm. Even within conventional mentorship – which relies on hierarchies, traditional power structures, and "best practices" – there is no guarantee that all will be well. In fact, it is often those very structures that give a sense of rigidity and false stability in the mentoring process (e.g., "If I do this, then ____ will necessarily happen"), producing instability that might create turmoil when things do not go as planned. Therefore, it is not that ruinous mentorship carries quantitatively *more* ethical considerations than other forms of mentorship; indeed, ethics are entwined with any and all mentorship. Instead, ruinous mentorship simply offers qualitatively *different* ethical considerations that are philosophically bound to our particular positions counter to humanism, convention, and a decaying planet.

From our positions, we have noted ruinous mentorship as a refusal as much as a disruption. It refuses humanist ideals; it refuses convention in methodology and mentoring; it refuses the status quo. Furthermore, it refuses the static and isolable subjects of humanism, disembodied from the material world, thinking themselves into existence like good Cartesian subjects. Refusing humanism means an ethical positioning that takes seriously the human as interconnected and converging with a dynamic and even vibrant (Bennett, 2010) world. Drawing from Braidotti (2019), our ethics is one of affirmation in that these connections become collectives that create possibilities for "constructing social horizons of hope" (p. 156). Through our refusals, then, we see the potential for hope and liberation. In other words, refusing to wallow in our shared humanist vulnerabilities, ruinous mentorship invites "radical relationality" (p. 166), in which we might expand our conception of mentors and mentees beyond its normative conception of two isolable bodies, instead thinking mentorship as a practice that always engages multiple others (e.g. other human and nonhuman bodies, other philosophical concepts, other forces). We think *differently*, relationally. Ruinous mentorship is, in effect, a type of community ethics. This affirmative and materially relational ethics cultivates different and differently responsible connections in the world, as well as in our mentoring practices.

Extending from our previous discussion on nostalgia, we reiterate our position that ruinous mentorship is, in fact, affirmative. Ruin begets hope, liberation, and even care. Ruining conventional mentorship means our responsibilities shift when we re-envision our relational commitments as we endeavor toward radical relationality. Our notion of care expands beyond any single one of us and ensnares us all as part of a ruinous mentoring assemblage. Braidotti (2019) reminds us that collectivization in the Anthropocene means that we can share the labors incited by the present and, for us, the labors in/with the ruins. Laboring and becoming affirmatively in the ruins means that we continue to seek ways to flatten hierarchies, to acknowledge difference, to make kin, so that care flows freely among us all. "Because affirmation is a practice that affects and transforms the negative conditions, and because of the infinite range of virtual actualizations, it is logically and materially impossible to exhaust all possibilities" (Braidotti, 2019, p. 180). Understanding both the ruins and ruinous mentorship in this way keeps us working toward more sustainable possibilities across our many projects. Indeed, ruinous mentorship might be the very thing that sustains us.

Conclusion: a dialogue

TRAVIS: *We focus on ruinous mentorship in our writing, but we know there are more than just that and conventional mentoring practices. We seek more to bring the vast possibilities of mentorship into awareness. Despite these new possibilities in mentorship, it feels difficult not to advocate for ruin as the way forward.*

KELLY: *Agreed. For both of us, ruinous mentorship is something that makes sense with our experiences as mentors/mentees. Certainly, there are times in my current position as someone who teaches qualitative methodology that, ethically, I realize that jumping right into ruinous mentorship might not work for the student. Per their desire, we find ourselves amidst a more conventional mentor/mentee relation. However, even when I find myself in those spaces, I often slip in moments of imbalance to unsettle things, like asking them for advice or sharing something I am struggling with in my scholarship. Sitting with a student who is looking to me as the "authority" makes me uncomfortable. I will intentionally challenge those hierarchies because I don't think they are productive, for either of us.*

TRAVIS: *That type of discomfort makes me feel like I could not refuse to be a ruinous mentor or constantly challenge their thinking, whether or not it was for their best interests. Perhaps I am too ruined to hold back.*

KELLY: *"Too ruined to hold back." I really like that. It's so interesting to think of the way ruin seeps through our thinking/doing – there's no going back or holding back. Maybe you are right. Maybe, even in the more conventional mentoring relations, ruin is always present.*

TRAVIS: *I think you are right in sensing ruin in all that we might think/do/ be. I am not sure we could ever really "know." I wonder if our mentees could refuse our efforts at ruination? As we refuse convention or tradition, could they refuse our refusals?*

KELLY: *Interesting questions. Refusing ruin; ruin as refusal. My inclination is to say "yes," they can refuse our refusals if they so desire. This is where we point back to ethics. We can only and always simply point to opportunities for ruin; however, whether or not they join us in the practices of ruinous mentorship is always up to them.*

TRAVIS: *And every intra-action leaves traces – they may refuse for now, they may embrace different possibilities; I guess we just continue to desire for them the liberation of ruin. It feels so humanist to think this way – that we know ruin, that we have the "choice" to refuse, that a mentee's act of refusing ruin is volitional. And yet, I cannot help but desire ruin for them despite my discomfort with it. Should we, even as ruinous mentors, desire something for others when we want them to be open to thoughts or becomings that we cannot yet envision?*

KELLY: *Traces, liberation, choice. I'm struck by all of these things, too. For me, the desire for ruin is something we can only offer, yet it is up to the other person to hail the call. That invitation is from where the ethical always flows, in that we can desire ruin, we can offer or invite ruin, but we cannot force ruin. They may not hear our call, or they can refuse it. Returning to the ethical, I actually think that the not-yet is where my optimism lies. You know me, always optimistic.*

TRAVIS: *I think that optimism is what I am feeling too, in my rather sundered way. If there is to be some coming to terms with the calamities in this epoch, it must rest on the not-yet. I guess desiring a world I cannot envision is a hopeful thought and one that ruinous mentorship may help to create. Even if there are no guarantees, the possibilities of ruin and the not-yet are something to strive for.*

KELLY: *So, you do have the potential for optimism after all! I, too, find hope in what's-to-come. It seems odd to say this, but ruin brings me hope. Ruin liberates us from the confines of convention, and it opens new possibilities for something different, something better we can create together.*

TRAVIS: *I am surprised by my optimism as well! I think refusing what we were "supposed" to be and refusing traditional mentoring practices to avoid humanist replication, to whatever degree, is vital to living in the Anthropocene – and perhaps living at all in the future. Perhaps only in these refusals can we be different, live differently; ruin may sensitize us to the possibilities of a more interconnected, intra-connected world.*

KELLY: *It all comes back to the material-relational. Bodies connected, bodies affected/affecting. If the Anthropocene teaches us anything, it is that humans are responsible for ruin, yet how we practice* *response-ability (Haraway, 2016) is still very much in our hands. You have often said, "the world is ending, no matter what we do." Perhaps you are right. However, how quickly it ends depends on how we act/enact. Whereas ecological ruin has gotten us into this mess, it is not predetermined where we will go from here. This is the same in our inquiry and our mentorship. We may find "ourselves" amidst the ruins of humanism; however, it is what we do in and with those ruins that comes to matter. Thus, we remain optimistic because ruin creates openings for possibility – in our world, in our inquiry, and in our mentorship.*

References

Achebe, C. (1994). *Things fall apart*. Penguin Books.

APA Presidential Taskforce. (2006). *Introduction to mentoring: A guide for mentors and mentees*. American Psychological Association. www.apa.org/education/grad/mentoring

Barad, K. (2011). Nature's queer performativity. *Qui Parle: Critical Humanities and Social Sciences, 19*(2), 121–158. https://doi.org/10.5250/quiparle.19.2.0121

Bennett, J. (2010). *Vibrant matter: A political ecology of things*. Duke University Press. https://doi.org/10.1215/9780822391623

Braidotti, R. (2006). *Transpositions: On nomadic ethics*. Polity Press and Blackwell.

Braidotti, R. (2011). *Nomadic subjects: Embodiment and sexual difference in contemporary feminist theory* (2nd ed.). Columbia University Press.

Braidotti, R. (2013). *The posthuman*. Polity Press.

Braidotti, R. (2019). *Posthuman knowledge*. Polity Press.

Butler, J. (1997). *The psychic life of power: Theories in subjection*. Stanford University Press.

Cartwright, N. (2007). Are RCTs the gold standard? *BioSocieties, 2*(1), 11–20. https://doi.org/10.1017/S1745855207005029

Deleuze, G., & Guattari, F. (1987). *A thousand plateaus: Capitalism and schizophrenia* (B. Massumi, Trans.). University of Minnesota Press. (Original work published 1980)

Grande, S. (2018). Refusing the university. In E. Tuck & K. W. Yang (Eds.), *Towards what justice? Describing diverse dreams of justice in education* (pp. 47–65). Routledge. https://doi.org/10.4324/9781351240932

Grossman, J., & Mackenzie, F. J. (2005). The randomized controlled trial: Gold standard, or merely standard? *Perspectives in Biology and Medicine, 48*, 516–534. https://doi.org/10.1353/pbm.2005.0092

Haraway, D. (2016). *Staying with the trouble: Making kin in the Chthulucene*. Duke University Press. https://doi.org/10.1215/9780822373780

Kuntz, A. M. (2015). *The responsible methodologist: Inquiry, truth-telling, and social justice*. Left Coast Press. https://doi.org/10.4324/9781315417332

Lather, P. (1997). Drawing the line at angels: Working the ruins of feminist ethnography. *International Journal of Qualitative Studies in Education, 10*(3), 285–304. https://doi.org/10.1080/095183997237124

MacLure, M. (2011). Qualitative inquiry: Where are the ruins? *Qualitative Inquiry, 17*(10), 997–1005. https://doi.org/10.1177/1077800411423198

Marn, T. M. (2018). *Performing the black-white biracial identity: The material, discursive, and psychological components of subject formation* (Unpublished doctoral dissertation), University of South Florida, Tampa.

McCoy, K. (1996). *Inquiry among the ruins: (Un)Common practices.* Paper presentation, American Educational Research Association (AERA) annual conference, New York City, NY.

Nick, J. M., Delahoyde, T. M., Prato, D. D., Mitchell, C., Ortiz, J., Ottley, C., Young, P., Cannon, S. B., Lasater, K., Reising, D., & Siktberg, L. (2012). Best practices in academic mentoring: A model for excellence. *Nursing Research & Practice, 2012,* 937906. https://doi.org/10.1155/2012/937906

Pickering, A. (1995). *The mangle of practice: Time, agency, and science.* University of Chicago Press.

Rackham Graduate School. (2015). *How to mentor graduate students: A guide for faculty.* Rackham Graduate School, University of Michigan. www.ucdenver.edu/docs/librariesprovider149/default-document-library/how-to-mentor-graduate-students.pdf?sfvrsn=3a6637b9_2

Rackham Graduate School. (2020). *How to mentor students: A guide for faculty.* University of Michigan. https://rackham.umich.edu/wp-content/uploads/2020/01/Fmentor.pdf

Somerville, M. (2016). The post-human I: Encountering "data" in new materialism. *International Journal of Qualitative Studies in Education, 29,* 1161–1172. https://doi.org/10.1080/09518398.2016.1201611

St. Pierre, E. A. (2014). A brief and personal history of post qualitative research toward "post inquiry." *JCT: Journal of Curriculum Theorizing, 30*(2), 2–19.

St. Pierre, E. A., & Pillow, W. (2000). *Working the ruins.* Routledge. https://doi.org/10.4324/9780203902257

Ulmer, J. B. (2017). Posthumanism as research methodology: Inquiry in the Anthropocene. *International Journal of Qualitative Studies in Education, 30*(9), 832–848. https://doi.org/10.1080/09518398.2017.1336806

Wolgemuth, J. R., & Donohue, R. (2006). Toward an inquiry of discomfort: Guiding transformation in "emancipatory" narrative research. *Qualitative Inquiry, 12*(5), 1012–1021. https://doi.org/10.1177/1077800406288629

THE VOICE OF CALM

Missy Springsteen–Haupt

Was my confidence foolish?
Did I used to feel bold?
Am I here by accident?

> *Your brain is working so hard.*
> *All of these experiences will help you.*

Stories aren't research
Teachers aren't systems leaders
Maybe you belong in a different program

What are numbers without stories?
What are leaders without teachers?

> *Don't let anyone tell your narrative.*
> *You are dismissed because they don't understand what you are saying.*
> *Your professional life is yours.*

It is not appropriate for researchers to be part of their studies
It can't be too personal
You need more distance

Is trading self the price of research?
If I disappear will my words be more true?

> *You are looking at things with the eyes of a researcher.*
> *Everything you are doing is feeding into who you want to be as an academic.*
> *Say what you need to say from your perspective.*

Am I moving too fast?
Am I moving at all?

> *Academia pretends to be in a hurry but it's a snail's pace.*

You bring clarity to chaos,
You open pathways I have been too afraid to consider
How can I thank you enough?

> *This is not a favor.*
> *I want to work with you.*

I admire you.

> *I admire you.*

Author Note: This poem for three voices represents her insecurities and frustrations (left), external and internal discouragement (middle), and the encouragement of her mentor (right). All mentor phrases are taken directly from conversations, text messages, and emails from her mentor, Dr. Amanda Jo Cordova at North Dakota State University.

LEARNING IS WATER REFLECTION

Cristina Valencia Mazzanti

Image 4.1 Learning Is Water Reflection

Source: Cristina Valencia Mazzanti

DOI: 10.4324/9781003022558-11

5

UNFINISHED

(POST-)PHILOSOPHICALLY INFORMED MENTORING AND RELATIONAL ETHICS

Candace R. Kuby

This chapter is a slowing-down, a close-reading of two books in relation to mentoring. The books are *The Incorporeal: Ontology, Ethics, and the Limits of Materialism* (2017) by Elizabeth Grosz, and *Unfinished: The Anthropology of Becoming* (2017) by João Biehl and Peter Locke. I put ideas from these readings in conversation with mentoring experiences as a way to think about be(com)ing a mentor. These readings speak to an ethics of now or in-the-moment relationalities, uncertainty and the unknown, nonlinearity of time and spaces, and the ways we are always, already intervening with/in/for each other. I was reading these books at the same time the editors of this book invited me to be a part of a panel at the International Congress of Qualitative Inquiry (ICQI) on mentoring in post-qualitative inquiry. I found my thinking provoked by these readings for many reasons, one being in relation to mentoring. These readings and my experiences shape how I think of the relational ethics between myself, students, the politics of inquiry, the academy, the communities we research with, and the lively world we are a part of becoming. Thus, I invited myself to think with these two books *and* mentoring, to see what might be produced for my thinking and the mentoring relationships I am a part of. What new ways of be(com)ing a mentor might come from this thinking?

 DOI: 10.4324/9781003022558-12

St. Pierre et al. (2016) write about newness:

> Whether work is "new" is always a matter of debate, and scholars
> doing new empirical, new material work usually begin by address-
> ing that issue and pointing out that the descriptor "new" does not
> necessarily announce something new but serves as an alert that we
> are determined to try to think differently.

(p. 100)

So how might I offer something that is new, in the sense it might help us to
try to think (and do mentoring) differently? What might my writing about
mentoring and these two books offer you? How might you, as a reader,
think with me and see what these books produce for you in mentoring relationships? This edited collection aims to explore the question: How is mentoring a philosophical practice? In this chapter, I specifically consider the
questions: What philosophies guide my mentoring as a relational practice?
How is mentoring a relation that affects *both* mentor (me) *and* mentees (my
students), flowing across, between, and together? And, what manifests or is
produced from mentoring relationships?

To situate my writing/thinking as I entered this chapter, I start with a
little about my research collaborations. My research in pedagogies of early
literacies (e.g., Kuby & Gutshall Rucker, 2016, 2020) and in pedagogies of
qualitative inquiry in higher education (e.g., Kuby & Christ, 2018, 2019,
2020a, 2020b) are inspired by post-philosophical concepts. For example, in
inquiring about young children's writing practices, I think-with Deleuze
and Guattari's (1980/1987) post-structural scholarship to consider the rhizomatic ways literacies come into being; the smooth and striated spaces of
pedagogies; and literacy desirings produced in relationships of students,
teachers, curricular materials, policies, art supplies, digital tools, and so
forth. My co-author, Tara Gutshall Rucker, and I also think-with writings on posthumanism (Barad, 2007), vibrant matter (Bennett, 2010), and
critical posthumanisms (Braidotti, 2013, 2018) to inquire into how literacy artifacts come to be. Similarly, in higher education, Christ and I use
post-philosophical ideas to inquire into *how* (pedagogies) we teach students to become qualitative inquirers, not just *what* (curriculum) we teach.
Inspired by post-philosophical ideas, we've written about speculative pedagogies of qualitative inquiry (Kuby & Christ, 2020a). In addition, I've

co-written a manifesto for teaching qualitative inquiry as arting, sciencing, and philosophizing inspired by Deleuze and Guattari's (1991/1994) writing on art, science, and philosophy as the three powers of thinking (see Kuby & Aguayo, 2019). These research endeavors over the years have prompted me to (re)think my beliefs on identities, subjectivity, and most significantly, ethico-onto-epistemology, or how we come to be/know/do in a lively, relational world.

Elizabeth Grosz (2017) starts her book with an introduction that states it is a book on ethics and ontology, specifically on materialism and idealism or this conception of bodies *and* minds. She spends one chapter each on a variety of philosophers, some I expected in the book (e.g., Spinoza, Nietzsche, Deleuze) and others that were new to me (e.g., Simondon and Ruyer). João Biehl and Peter Locke (2017) co-edit a book that conceptualizes anthropological research – often framed as ethnographies – with philosophical ideas informed by Deleuze and Guattari's writing on becoming. The collection of chapters span many international contexts and projects. As I read the introductions in both books, I found myself highlighting, responding to, and marking up more of the pages than not. There was so much to think with! I thought it would be a fruitful exercise to read these books through/with my experiences as a mentor and being mentored by others and see what I might offer readers to think-with as well.

After more than 15 years into my teaching in higher education, I wanted to slow down and think with and about mentoring philosophically inspired by post-philosophical concepts. Inspired by other qualitative inquirers, slowing down, as an ontological (and axiological and epistemological) practice is much needed for me and perhaps you. As Ulmer (2017) poses in her article titled, "Writing Slow Ontology," what if we as scholars create writing that is not unproductive but is differently productive? She articulates this as a space of creative intervention in the escalating pace of academic production. I think this notion of slowness can also be thought in relation to the pressures and accelerated pace scholars might feel to mentor others in the "right" way as deemed by bodies of literature on mentoring. Bozalek (2017) and Leibowitz and Bozalek (2018) write of slow scholarship in relation to writing and writing retreats in academia and invite us into "response-able" pedagogies in our teaching as a way of enacting slow scholarship. Could we also use creative interventions or response-able pedagogies as a way of slowing down to think-with post-philosophies in an effort to be(come) differently productive in our mentoring relationships?

Being tasked with the otherwise: mentoring improvisation

I start with some lines from the opening two pages of the introduction in Grosz's (2017) book:

> We create ontologies.
> How to bring about a future different from the present.
> Possibility of the future being otherwise than the present.
>
> (pp. 1–2)

As mentors, we enter into mentoring relationships with our past experiences as students, current expectations at our institutions regarding coursework and mentoring, and the students we engage in relationships with. Likewise, these students have past and present experiences that are all a part of the moments we find ourselves in with them. We – do create ontologies. But our work as mentor/mentees is already entangled with the future. Or, said another way, we engage with a nonlinearity of time and spaces. We have a responsibility to consider that how we mentor is the future we inherit in the academy. Or, as Barad (2013) writes, "we inherit the future" (p. 23). Past, present, future collapsed in mentoring relationships. How might the future be otherwise? Or perhaps, as Grosz (2017) writes, what are "possibilities for being otherwise?" (p. 13).

In Biehl and Locke's (2017) opening pages of their foreword, they invite readers into the book with a series of questions:

- How can we ethnographically apprehend these worldly fabrications and the lives therein, constituted as they are by that which is unresolved, and bring this unfinishedness into our storytelling?
- How are long-standing theoretical approaches able – or not – to illuminate emergent political, economic, and affective realities?
- How can the becomings of our informants and collaborators, and the movements and counterknowledges they fashion, serve as alternative figures of thought that might animate comparative work, political critique, and anthropology to come? (p. x)

Biehl and Locke posit that "ethnographic creations are about the plasticity and unfinishedness of human subjects and lifeworlds" (p. x). Might that be a provocation for us to think with mentoring? Mentoring relationships are about plasticity and unfinishedness. What might these concepts produce for us in how we go about mentoring relationships? Biehl and Locke go on to write about ethnographic work, but perhaps we can substitute the phrase "mentoring work":

> ethnographic [mentoring] work always begins in the midst of social life, its rhythms, affects, surprises, and urgencies. . . . Becoming troubles and exceeds our ways of knowing and acting. It pushes us to think against the grain, to consider the uncertain and unexpected in the world, and to care for the as-yet-unthought that interrogates history and keeps modes of existence open to improvisation. We are tasked with the otherwise.
>
> (p. x)

So how might this inspire our mentoring relationships? To be tasked with the otherwise. To embrace uncertain and unexpected ways of being and doing mentoring. To improvise rather than to solely "know" how to mentor or to rely solely on information given to us about what makes a "good" mentor. Biehl and Locke continue to write about ethnography with a call to the singular, micro, and partial, instead of repetition and models of perhaps the "tried and true ways." I also wonder if this might mean for us to let go of or question the "tried and true ways" to be a "good" mentor. Perhaps mentoring is about "self-world entanglements and leaking social fields" (Biehl & Locke, 2017, p. x) rather than a checklist or pre-set formula, or perhaps a both/and.

Be(com)ing together: mentoring as collection relations

Mentoring
verb
Advise or train (someone, especially a younger colleague).
(Google dictionary definition)

As I sat on my screened-in porch thinking, reading these books, and writing the conference presentation that became this book chapter, I felt stuck. On a recent flight, I had re-read aspects of the Grosz book, highlighted parts I felt spoke to this topic of philosophically inspired mentoring, and even made notes on an index card. But several days later as I sat down to write, the connections and big ideas seemed to evade me. What do I have to say about mentoring? What newness might I have to offer? I then remembered the quote by St. Pierre et al. (2016) I shared above about newness. How might I slow down my thinking or relationship with the word "mentoring," not in an effort to nail-down the "right" definition, but in the process of thinking-with-mentoring be provoked to think otherwise? This practice of musing on definitions and etymologies is "not in a quest to find 'The' answer but rather it is [in] a process of searching for definitions and etymologies that our thinking is undone and newness about a concept/word is produced" (Kuby & Christ, 2020a, p. 12).

Thus, after I looked up the dictionary definition for mentoring, I thought back to specific students who I have mentored in official capacities on dissertation committees or through courses I taught. I also thought about someone who I see as my mentor, Dr. Jennifer Rowsell, and what she does – her ways of knowing, being, and doing – that have been (and continue to be) so inspirational to me. I went back to a letter I wrote for Jennifer when someone else was putting together an awards packet for a mentoring award. I asked her permission to share an abridged version:

> She [Jennifer] gives such specific feedback framed in a way that is careful and thoughtful. She encouraged me to work towards publication as a doctoral student (read drafts and gave suggestions for journals even though I was not a student of hers, she was not an official mentor of mine from my doctoral granting institution). This speaks to her generous heart for young scholars in the field of educational research. She sees hope in the newer generations and is excited to mentor numerous scholars that are *and* aren't affiliated with her university and official job duties.
>
> She worked with me to navigate various journal options, put me in contact with editors, and helped to draft a proposal for a special issue of a journal. Once abstracts, and then manuscripts came in, she was influential in helping me understand an editor's role and how to communicate ethically with reviewers and authors. In this process of co-editing, I didn't feel belittled as an early career

scholar but rather as an equal, someone with expertise, and capable of learning how to be an editor. This speaks to her grace and ethical ways of mentoring. Reflecting on my experiences with her have definitely shaped how I have begun to mentor new faculty members as well.

She has been intentional to invite me to attend "by invitation only" international workshops and/or publish in edited books and handbooks that she is leading. The professional networking at these workshops and the wide audience of readers to the books she edits has been significant in my early career years.

She understands that mentoring is about helping newer scholars make connections with people locally but also in wider international communities. She understands the politics of tenure and promotion and offers, explicitly, to nominate me for any awards that are a good fit for my scholarship.

However, my story with Jennifer is not unique. Many others have shared with me similar sentiments and specific ways that this scholar has mentored them. She leads by example and is able to give specific advice on the process of becoming a new faculty member. Her ethics, humble heart, and genuine positive energy are contagious. She is truly a gem in the educational research community.

Perhaps these excerpts from the letter connect to your own experiences as a mentor or someone being mentored. I hope so. There is something about my relationality to *and* with *and* from Jennifer. I can't really explain it or put it into words. I can't even trace how it came to be. Our first interactions were happenstance or a thrown-togetherness at a working conference where we happened to be placed in the same small group. This reminds me of the Biehl and Locke quote presented earlier and my reading of it as related to mentoring in the midst of social life, its rhythms, affects, surprises. The relationship with Jennifer was a surprise, an unexpected relationship born in the midst of social life. Over the years we've found rhythms together and ways of being response-able with/to each other. There is an easiness to conversation, a sincerity not only about work but also our families; there is a deep ethics in what and how we do our work as scholars, a deep ethical way of being and becoming together. We are always, already intervening in/with/for each other. Or, as Biehl and Locke (2017) write, my relationship with Jennifer feels as a "as-yet-unthought . . . that keeps modes of existence open to improvisation" (p. x). Or, as Grosz (2017) writes, "Bodies are bodies only to the extent that they are capable of acting on each other through collective relations" (p. 27). Over time, we've found that we need thinking-collaborating projects going between us and miss the times that we don't. We both have stated we need collaborative projects as we find our moments together jointly productive in shaping who we are and the scholarship we engage with.

Mentoring is/as collective relations. Bodies acting on each other. A becoming-with or becoming-together. As Jennifer shares with me, she learns with me and seeks out opportunities for us to collaborate together. It isn't a hierarchical feel to mentoring (one person is the mentor and the other is the mentee); or, said another way, a binary relation as the definition of the word suggests earlier, but rather a becoming relationship. Thus, the hierarchy and/or binary mentor/mentee relationship feels fuzzy and not predetermined. And while it doesn't *feel* hierarchical, there are identities at play such as gender (we both identify as female), race (both identify as white), and rank (Jennifer was already an established scholar whereas I was a doctoral student when we met). Within one conversation, I feel myself moving in and out of mentor *and* mentee with Jennifer. I would suspect she'd say the same.

So thus, while I don't experience our relationship as hierarchical, the aforementioned letter points to the hierarchy that does exist in the academy in how I felt I should write about her as a mentor: the specific things she does, valued in the academy, of an established scholar with a newer scholar. So, while I want this letter to demonstrate a relationship that feels different to me from other mentoring relationships I've been a part of, it also functions as evidence of "good" mentoring for a neoliberal academy. A both/and. So how do letters such as this one, for a mentoring award, reinscribe a version of mentorship that is linear and binaried? How might we write this letter otherwise? How might we write this letter through Grosz and Biehl and Locke?

Biehl and Locke (2017) write of becoming as, "Always singular yet ever producing multiplicity, the work of becoming is inherently a work of creation" (p. 9). To think of mentoring relationships as a work of creation prompts me to be open and not have a fixed agenda and/or plan to follow linearly with each person I engage in mentoring with. Each mentoring relation creates its own multiplicity. Biehl and Locke go on to write, "Becoming is always excessive and unruly – to be unformed is a kind of active (if unanticipated) resistance" (p. 11). How do Jennifer and I resist hierarchical and binaried ways of mentoring yet at the same time work within those expectations and structures? Be(com)ing both/and? Grosz (2017) asks:

> How can one live in the world, act in the world, make things and oneself, while also creating values that enhance oneself and one's milieu, not through preexisting values but through acting, making, and doing that generate new values? There is no single answer to this question, only ways or styles of living in their multiplicity.
>
> (p. 255)

Generating new values. Acting. Making. Mentoring as an active (unanticipated) resistance: this is something I invite us to ponder more. What if we don't think of mentoring as training or advising from a checklist? What might the active, yet unanticipated, resistance produce for us in mentoring relations, the academy, and communities we research within?

An ethics of the now: mentoring as an active (unanticipated) resistance

This be(com)ing-together, active unanticipated resistance, demands a rethinking of ethics – not just what we do as mentors (e.g. the checklist, list of conversations and events our institutions suggest we engage with when we mentor) but rather an in-the-moment ethics. I think back to a moment when then doctoral student Oona Fontanella-Nothom moved cross-country with her family from California to Missouri to begin a doctoral program with me as her advisor. When they arrived from several days of driving, the home they were supposed to rent was not inhabitable for her family. In a new city, with only the belongings in their car, with a moving truck on the way, she wasn't sure what to do. In that moment, I thought, yet didn't think of what to do. I invited her, her husband, and son to live with me until they could secure a better place to live – to take over the bedroom I share with my husband. Was this the "right" way to be an advisor? Was this something the university and/or my department would support? In the moment, my decision (with someone who I had never met in person before) was to open my home to them. She often recounts this story – the moment I literally opened my front door to her family for the first time – the moment of becoming-mentor-mentee in her narratives about graduate school. I realize that not all mentors might have the resources and/or physical space to respond the way I did to Oona; however, that is not my point. Rather, I want us to consider the relational, in-the-moment ways we are already, always be(com)ing mentor(ed). How everyday, perhaps sometimes mundane, actions create ontologies and relationships. For me, I didn't have to think about whether to open my home or not; it was a given in my mind and heart, yet it was a pivotal moment for Oona and her family.

Mentoring as an ethics of the now. As Grosz (2017) writes:

> a way of thinking about not just how the world is but how it could be, how it is open to change, and above all, the becomings it may undergo. In this sense, an ethics always passes into and cannot be readily separated from a politics, which addresses social, collective, cultural, and economic life and their possibilities for change . . . Yet an ontology entails a consideration of the future, not only of what we can guarantee or be certain but above all what virtualities in the present may enable in the future. This is the possibility of the future being otherwise than the present, the openness of a future which is nevertheless tied to, based on but not entirely limited by, the past and present.
>
> (pp. 1–2)

So how do we think-about and think-with our mentoring relationships – the past, present, futureness of them – the in-the-moment ethics? Grosz (2017) writes "the question [for her book is] of how to act in the present and, primarily, how to bring about a future different from the present" (p. 1). How do we bring about a future different from the present in the academy through our mentoring relationships and what seems like the mundane everydayness of the academy? This connects to Barad's quote, stated earlier, that we inherit the future. Or more specifically related to this book's focus, what does the mentoring of qualitative inquiry look like in the academy? And how is qualitative inquiry mentoring in the present (and in the past) the future we inherit?

As a mentor, I have often been surprised when students share things about my mentoring and teaching that stick with them or that were significant in their learning journeys. For example, students in my qualitative inquiry courses share what my syllabus has done or produced for them (Kuby & Christ, 2020a). A syllabus seems like a mundane, everyday, or expected item in a college class. Yet, I've had to slow down and think about what syllabi do or produce for students and our relationships as learners together. Some students talk about my language in qualitative inquiry class sessions, when I use phrases such as "it depends" or "give yourself permissions to _____" (Christ & Kuby, 2020; Kuby & Christ, 2018, 2020b). While I know language is important in teaching, it wasn't until I had Becky Christ join my classes weekly as a teaching intern/research assistant that I became aware of specific phrases I used, and we began to see what these produced. Other students share about moments during classes when I invite students to think/learn through making with artistic tools (Kuby & Christ, 2019, 2020a). While artistic tools in an introductory qualitative inquiry course might not seem mundane and everyday, they are tools in a general sense that are mundane – crayons, glue sticks, markers, colored paper. However, in the relationalities of a doctoral class on the doing of qualitative inquiry, they produced different and new ways of knowing/doing/be(com)ing.

As I think of mentoring as unanticipated resistance, I think to another letter. The following is a quote from a letter three doctoral students wrote about my relationship with them as a mentor and teacher in several qualitative inquiry courses:

> as a mentor, Dr. Kuby supports the process of allowing oneself to be uncomfortable and move out of their comfort zone. Dr. Kuby offers opportunities for students to experience disequilibrium, becoming intellectually stuck and thus, embracing the journey of digging a way out. As a result, she averts providing clear answers [to] student theoretical inquiries directly, instead offering a slow pause where she responds inquisitively with, "Tell me more." In these intentional pedagogical choices, Dr. Kuby makes room for multiple perspectives to be heard, considered, and valued. Dr. Kuby's courses create space for creativity in thinking and course products, such as writing, presentations, and visual imagery. Dr. Kuby's pedagogy embraces creativity, disrupting the normative and commonsensical practices often present in higher education classrooms. We feel that Dr. Kuby's pedagogical and mentoring abilities have assisted us in

building confidence in ourselves as risk-taking emerging scholars, writers, and researchers. . . .

As a mentor and advisor, Dr. Kuby is known around campus as a faculty member you can rely on to provide consistent support, feedback, and strong advising. We respect her mentoring, advising, and teaching that comes with not only high expectations, but also a disciplined and determined attentiveness that encourages students to follow their inquiry curiosities wherever they may lead, while embracing the discomfort of not knowing in the midst of continually growing and evolving as scholars.

(letter written by Oona Fontanella-Nothom,
Palwasha Khan Marwat, and Traci Wilson-Kleekamp)

So, similarly to the earlier letter I wrote for Jennifer, this letter excerpt from my students could be read hierarchically or in the binary of mentor/mentees. However, the examples they discuss illuminate how everyday, perhaps mundane, interactions and decisions produce ontologies and ways of knowing/be(com)ing/doing research for the students we mentor and teach. Or as Grosz (2017) writes, as quoted earlier, "Yet an ontology entails a consideration of the future, not only of what we can guarantee or be certain but above all what virtualities in the present may enable in the future" (pp. 1–2). What virtualities in our mentoring in the present enable the future? How can our mentoring be an active (unanticipated) resistance to the way things have always been? To the hierarchical and perhaps even expected ways of mentoring in the academy? These questions beg us to consider in-the-moment ethics. As Grosz (2017) writes, "An ethics does not spring directly from our understanding of the world. Rather, it comes from our affective bonds to and connections with other things in the world, relations that enable us to enhance [and/]or diminish forms of life" (p. 7). Our daily, affective bonds or connections with those we are engaging with/in mentoring. Thus, we are tasked with thinking/doing/knowing/be(com)ing mentoring otherwise. Post-philosophical concepts force us to (re)think mentoring and how we engage with it in our daily, mundane everydayness.

Similarly to my interactions with Jennifer, over time my relationships with Oona, Palwasha, and Traci, have provided spaces where I feel mentored by *them*. A both/and. These women identify in a variety of ways related to race and ethnicities. However, we all four use the same gender pronouns, and we acknowledge the varied historical and present contexts and experiences that shape how we engage within the academy and with each other. They each have offered much to me, to learn about their world experiences, which shapes how I live/teach/think. So, while I engage in expected mentoring relationships such as providing research assistantships, inviting students to co-author manuscripts with me, or serve as a teaching intern in courses I teach, we also have other ways of being that shape the mentoring relationships as collective relations of otherwise. For example, we have conversations about international politics, racism and policies and practices in our city and state, and difficult times in our personal lives with families and friends. We create spaces to enter into relation with each other, or to make kin, as Donna Haraway (2016) might say. We provide meals for each other when someone has a surgery (or a family member); we swap food from our gardens; we invite one another over for meals and holiday gatherings. We engage with matters of care or world-making a/effects with/ for each other as María Puig de la Bellacasa (2017) discusses so eloquently in her book about speculative ethics in more than human worlds.

Making a world with/in/through mentoring

If we think back to an earlier quote by Grosz, she states, "We create ontologies." She goes on to write "Ontologies have ethical and political implications in the sense that they make a difference to how we live and act, what we value, and how we produce and create" (Grosz, 2017, pp. 3–4). Similarly, Biehl and Locke (2017) write about becoming and ethics as "conditions under which something new might be produced" (p. 7). So as I end my musings, provocations, and thinkings about mentoring, I invite you to consider the mentoring relationships you are a part of. I don't have answers or a clearcut checklist on what to do next that works with the lively ontologies that these authors are inviting us to think about. Rather, I leave you with the following quote to ponder, linger, and sit with. Biehl and Locke (2017) write:

> To draw on this productive unmooring, we might need to let go of some venerable assumptions about the human condition and about where we locate political action, instead of asking what life-forms, collectives, and new kinds of politics are on the horizon, brewing within the leaking excesses of existing force fields and imaginaries. Becoming, thus, is a style of noticing, thinking, and writing through

which to capture the intricate relations, movements, and dynamics of power and flight that make up our social worlds.

(pp. 9–10)

What assumptions about the human condition might we, might you, might the academy, be making in relation to mentoring? And as Biehl and Locke write, what might we need to let go of? We exist and live in a lively, relational world of both/ands; however, we have created a falsely binaried world of identities, labels, categories, checklists, hierarchies, and ways of be(com)ing/knowing/doing. If we slow down, if we think our world as lively collective relations, if we think differently to produce a differently or otherwise world, what might mentoring become? What might mentoring do? What future might we inherit? Biehl and Locke (2017) explain that becoming is always between or among – or said another way, becoming "makes a world" (p. 14). How do we make a world with/in/through mentoring?

Acknowledgments

Thank you to Jennifer, Oona, Palwasha, and Traci for your permission to include excerpts from letters in this chapter and to include your names. I appreciate the feedback shared with me on drafts of this chapter, and most of all, I cherish our mentoring relationships with each other.

References

Barad, K. (2007). *Meeting the universe halfway: Quantum physics and the entanglement of matter and meaning.* Duke University Press. https://doi.org/10.1215/9780822388128

Barad, K. (2013). Ma(r)king time: Material entanglements and re-memberings: Cutting together-apart. In P. R. Carlile, D. Nicolini, A. Langley, & H. Tsoukas (Eds.), *How matter matters: Objects, artifacts, and materiality in organization studies* (pp. 16–31). Oxford University Press. https://doi.org/10.1093/acprof:oso/9780199671533.001.0001

Bennett, J. (2010). *Vibrant matter: A political ecology of things.* Duke University Press. https://doi.org/10.1215/9780822391623

Biehl, J., & Locke, P. (Eds.). (2017). *Unfinished: The anthropology of becoming.* Duke University Press. https://doi.org/10.1215/9780822372455

Bozalek, V. (2017). Slow scholarship in writing retreats: A diffractive methodology for response-able pedagogies. *South African Journal of Higher Education, 31*(2), 40–57. https://doi.org/10.20853/31-2-1344

Braidotti, R. (2013). *The posthuman.* Polity Press.

Braidotti, R. (2018). A theoretical framework for the critical posthumanities. *Theory, Culture, & Society, 36*(6), 31–61. https://doi.org/10.1177/0263276418771486

Christ, R. C., & Kuby, C. R. (2020). Speculative (wombing) pedagogies: |Rə'zistəns| in qualitative inquiry. In N. Denzin & J. Salvo (Eds.), *New directions in theorizing qualitative inquiry: Theory as resistance* (pp. 24–38). Myers Education Press.

Deleuze, G., & Guattari, F. (1987). *A thousand plateaus: Capitalism and schizophrenia* (B. Massumi, Trans.). University of Minnesota Press. (Original work published 1980).

Deleuze, G., & Guattari, F. (1994). *What is philosophy?* (H. Tomlinson & G. Burchell, Trans.). Columbia University Press. (Original work published 1991)

Grosz, E. (2017). *The incorporeal: Ontology, ethics, and the limits of materialism.* Columbia University Press.

Haraway, D. J. (2016). *Staying with the trouble: Making kin in the Chthulucene.* Duke University Press. https://doi.org/10.1215/9780822373780

Kuby, C. R., & Aguayo, D. (2019). A manifesto for teaching qualitative inquiry with/as/for art, science, and philosophy. In C. A. Taylor & A. Bayley (Eds.), *Posthumanism and higher education: Reimagining pedagogy, practice and research* (pp. 73–83). Palgrave Macmillan. https://doi.org/10.1007/9783030146726

Kuby, C. R., & Christ, R. C. (2018). Productive aporias and inten(t/s)ionalities of paradigming: Spacetimematterings in an introductory qualitative research course. *Qualitative Inquiry, 24*(4), 293–304. https://doi.org/10.1177/1077800416684870

Kuby, C. R., & Christ, R. C. (2019). Us-ing: Producing qualitative inquiry pedagogies with/in lively packets of relations. *Qualitative Inquiry, 25*(9–10), 965–978. https://doi.org/10.1177/1077800419843563

Kuby, C. R., & Christ, R. C. (2020a). *Speculative pedagogies of qualitative inquiry* (Research Methods Series). Routledge. https://doi.org/10.4324/9780429285707

Kuby, C. R., & Christ, R. C. (2020b). The matter we teach with matters: Teaching with theory, theorizing with (textbook) bodies. *Qualitative Inquiry, 26*(1), 71–80. https://doi.org/10.1177/1077800419874826

Kuby, C. R., & Gutshall Rucker, T. (2016). *Go be a writer! Expanding the curricular boundaries of literacy learning with children* (Language and Literacy Series). Teachers College Press.

Kuby, C. R., & Gutshall Rucker, T. (2020). (Re)thinking children as fully (in)human and literacies as otherwise through (re)etymologizing intervene and inequality. *Journal of Early Childhood Literacy, 20*(1), 13–43. https://doi.org/10.1177/1468798420904774

Leibowitz, B., & Bozalek, V. (2018). Towards a slow scholarship of teaching and learning in the South. *Teaching in Higher Education, 23*(8), 981–994. https://doi.org/10.1080/13562517.2018.1452730

Puig de la Bellacasa, M. (2017). *Matters of care: Speculative ethics in more than human worlds.* University of Minnesota Press.

St. Pierre, E. A., Jackson, A. Y., & Mazzei, L. (2016). New empiricisms and new materialisms: Conditions for new inquiry. *Cultural Studies-Critical Methodologies, 16*(2), 99–110. https://doi.org/10.1177/1532708616638694

Ulmer, J. B. (2017). Writing slow ontology. *Qualitative Inquiry, 23*(3), 201–211. https://doi.org/10.1177/1077800416643994

She said I should call her Jenni

Krista S. Mallo

The summer of two years ago – when my father had been gone already two years, when my husband was two years into a new career, when my older children were in the throes of adolescence, when my middle children were adjusting to being displaced by the baby who was six months old –

When I decided to take two more classes to see if I could possibly manage any kind of now-or-never forward motion in my doctoral program –

When we had outgrown our house and had the urgent opportunity to sell and move into a bigger house with enough bathrooms and square footage and a mortgage we might be able to afford –

When the professor of one class said no more excuses, no more extensions, no more questions, and no more leniency or grace – or mercy – and I don't even know what that professor's name was – the summer of two years ago –

When I fell so far behind and had been dropped so hard by No-Name-No-Grace-No-Mercy that I could barely breathe when the other – *the other* – asked for a phone call to discuss my late still missing work –

When my heart pounded migraine pain directly up my throat into my brain, swelling as salty tears often do –

I escaped from the three meltdowns happening inside the walls of my chaos, sat in my own minivan in my own driveway, windows up, air

DOI: 10.4324/9781003022558-13

conditioner on full blast, cold, so cold, trying to be stone cold, so numb, so that when –

> I answered when she called: "Yes, Doctor – . . ."
> *immediately, she interrupted . . .*
> "You should call me Jenni."

Before we go any further in this conversation, she seemed to say, *we will not make any progress if we are both caught up in formalities of institutional policy and procedure. Before we go any further in this conversation*, she seemed to say, *we will make progress only if we are on the same page, with the same goals, moving forward together, side by side*, she seemed to say. *I will not stand over you, she seemed to say, and grab hold of you and throw you over any cliff that you are already considering leaping headlong off,* she seemed to say.

> "There are some things you need to learn," she continued. "I really don't care how long it takes you to learn them, even if the semester ends, as long as you learn them."

It was a month or so ago at the culmination of my dissertation finally – when each member of the committee appeared on the unbelievably pandemic-proof virtual stage alongside me – with her as the defense chair –

As I began to speak, I could hear those words echo in my how-did-I-get-here memories – impact, affect – extended ripple effect – the summer of two years ago – when she became my mentor, not even sure she knows what happened when

she said I should call her Jenni.

MY NORTH STAR

Quintin R. Bostic II

Maya Angelou once said, "In order to be a mentor, and an effective one, one must care." In pursuing my PhD, I have learned that this level of care and mentorship is different for everyone, especially for us Black and Brown people. Historically, the North Star was used by slaves to escape to freedom. Miles Davis said, "Knowledge is freedom and ignorance is slavery." Like my ancestors, I used my own North Star, my mentor, during my PhD program to learn the skills needed to thrive as a Black academic. When entering the PhD program, I had no idea how challenging it would be for me as a Black man to navigate the world of higher education. I have experienced being told by White professors to use "African American" and not "Black" as it is more of an appropriate term to describe people who look like me. I have even been in a university classroom where a student and professor used the term "coon." The problem is that if these professors taught through a culturally responsive lens, they would know that "Black" is a commonly used term that many people of color identify by, and that the term "coon" is a racist word used for describing Black people. When culturally responsive mentoring for Black and Brown students is not embedded in the everyday interactions that mentors use to support their students, the outcome can be troublesome. Without my North Star, I would not be the fearless PhD candidate I am today. Receiving culturally responsive mentoring from her helped me to reach deep down in myself and make connections to the

DOI: 10.4324/9781003022558-14

academy that I did not see as being possible. My North Star helped me internalize and see the permanence of racism in society and how it exists and controls various social, political, and economic spaces, including higher education. My North Star helped me to understand and embrace the idea of being a voice of color to my White counterparts as my knowledge and experiences can challenge and transform racial dominance. She did this by engaging me in book studies and small group discussions surrounding critical race theory. I was included in her social network that was comprised of Black and Brown professors who shared the same challenging experiences that I face in the academy. These social connections were able to teach me things that textbooks were not able to. It is not enough to address challenges by reading books and research journals – it takes a person who has shared those same lived experiences to help you understand. The beauty of it all is that I see myself in my North Star – I see my race/ethnicity, my experience, my culture – everything that makes me who I am today. My North Star tapped into that, thus, leading to my success. For the sake of Black and Brown academics, culturally responsive mentoring must always be considered when mentoring students.

6

MENTORING AS RADICAL INTERCONNECTIVITY AND LOVE

POST-OPPOSITIONAL AND DE/COLONIAL APPROACHES TO QUALITATIVE RESEARCH MENTORING IN HIGHER EDUCATION

Kakali Bhattacharya and Jia Liang

In this chapter, we describe our philosophical orientations to mentoring using post-oppositionality (Keating, 2013) and de/coloniality (Bhattacharya, 2009). We write from the perspectives of two women of color in predominantly white teaching and learning spaces in higher education. Kakali, a professor of qualitative inquiry, describes her philosophical orientations in the recursive context of training students to engage in critical and de/colonial inquiry for over a decade (Bhattacharya, 2006, 2009; Denzin et al., 2008; Rhee & Subreenduth, 2006; Smith, 1999/2012). Jia, an assistant professor, describes her orientations and insights from bearing witness to and receiving Kakali's informal mentoring.

Informed by post-oppositionality (Keating, 2013), we have also accepted the invitation to move beyond binaried oppositional discourses that are incentivized in academia and embraced the idea of expanded awareness and broader possibilities. In doing so, we open up spaces for relational connectivity that could illuminate possibilities for reconfiguring teaching and learning qualitative research. This interconnectivity cultivates a sense of commitment in mentoring as we navigate the problematic terrain of academia and qualitative inquiry, which is dominated by whiteness-centering discourses. For example, most qualitative inquiry textbooks originate in

DOI: 10.4324/9781003022558-15

predominantly white countries and are written by white researchers. Whiteness is intimately connected with ways of knowing, doing, and teaching qualitative inquiry. Therefore, we come together, as non-white researchers, complicit with whiteness due to our training and academic positions, to share our perspectives and make space through mentoring to honor critical and de/colonial relational practices.

We share both risk-taking and risk-avoiding moves in our relational mentoring practices. The recursive nature of these practices allows Kakali to continuously carve out new methodological spaces for her mentees. In contrast, Jia must negotiate these spaces with caution within her predominantly white discipline, educational leadership, highlighting the possibilities and limits of critical methodological moves.

We discuss four mentoring practices grounded in the ontoepistemologies that inform our qualitative inquiry: (1) engaging in shadow work and inner journeying; (2) un/learning established approaches to qualitative inquiry and honoring culturally situated ontoepistemologies; (3) cultivating imagination, play, and whimsy; and (4) spacemaking for newcomers' raw intelligence and communal knowledge (i.e., theorizing from the street) (Bhattacharya, 2019). In what follows, Kakali explains the four practices and their manifestation in her mentoring. Jia describes the influence of these practices as their recipient. Together, we discuss how these practices influence our relationships, pedagogy, and critical and de/colonial qualitative approaches, fostering our understanding of mentoring as radical interconnectivity and an act of critical compassion and love.

Entering this work

From 2014 to 2018 we worked in the same department at the same institution. As the only women of color in the department, and having earned our doctoral degrees at the same institution, we quickly established a familiarity. It took longer to cultivate a trusting relationship, as both of us have histories of heartbreak and betrayal in academia. Kakali was tenured when we met and became a full professor in 2018. Jia went up for her pre-tenure third-year review in 2017 and received unanimous positive feedback.

Forming a relationship: being in community

We did not enter into a formal, institutionalized mentoring relationship. In fact, the institution had its own mentoring structure in which we both participated, but we were not paired through that mentoring program. Our relationship developed over months and years, gradually deepening, based on how we learned to trust and lean on each other as we both navigated hostile academic structures and agents who perpetuate them. At first, we were simply colleagues. Gradually, we tested each other to discern whether we could hold safe space for each other. Jia worked closely with Kakali to develop a qualitative research graduate certificate. Along the way, she was able to observe Kakali's navigation of the political power plays and successfully launch an 18-credit-hour qualitative research graduate certificate. Later, at a qualitative research software training retreat, we spent a lot of time together, away from our institution. This allowed us to engage in multiple, prolonged, honest, and vulnerable conversations, more than we could with our busy schedules at our institution. Once we returned to our institution, we stayed in dialogue and deepened our friendship. Jia witnessed multiple moments where Kakali took intellectual and professional risks, and attracted harm to herself, while advocating for Jia. After a while, it became a natural progression of our relationship, where we fell into a comfortable space of community building with each other.

Jia, through her experiences in the U.S., learned to navigate hostile, white academic spaces with a guarded nature, being keenly observant, and choosing strategic silence. Thus, if the relationship did not develop organically, Jia would have not been able to trust the strength of the relationship and remain unguarded. *Mentoring* is perhaps not the best word to describe the relationship. Instead, by being in community with each other, Kakali was able to leverage her positional privileges for advocacy, guidance, and disruption of oppressive structures in qualitative research. Jia became a thought-partner for Kakali for various academic and political issues and teaching qualitative research to a predominantly white audience who needed to interrogate their privileges critically.

We never really formally related to each other as a mentor and a mentee. In fact, we never had any intentional conversation of being in such a relationship even when we began to trust each other. However, we come from cultural traditions that honor the wisdom of those who have traveled before us on the path we are about to embark. We approach those relationships with humility, and it is alien to our cultural constructs to consider such relationships outside a learner-learned structure. This is not to mean that the learner is not learned or the learned has stopped learning. However, in this scenario, exposing what Kakali has learned would violate Jia's desire to remain guarded outside of her trusted spaces. Thus, a philosophical discussion of mentoring within the context of qualitative research is not devoid of the cultural implications. Additionally, our relationality is not informed by postmodernist ideas of fluidity and multidirectionality, although we align with those ideas in some of our published works. In this context, we relate to each other through a shared structure of respect, trust, friendship, solidarity, community, humility, and a desire to learn from wisdom and experience.

Kakali's narrative

I have taught qualitative methods since 2005. I received my PhD from the University of Georgia (UGA) in educational psychology with a specialization in qualitative inquiry. I was trained in qualitative methodology. I locate myself in de/colonial, transnational, feminist ontoepistemologies and approaches. I have defined my framing of de/coloniality in multiple spaces (Bhattacharya, 2009, 2015) situating de/coloniality as a movement between existing in resistance to colonizing forces and dreaming of being in relations and spaces that are devoid of colonialism. This imagination exceeds the restrictions of time and space. I bring this understanding of de/coloniality by putting my Global South sensibilities in dialogue with the work of Gloria Anzaldúa and AnaLouise Keating.

I discovered Anzaldúa's work during my graduate studies and have continued to immerse myself more deeply in her thought over the years. Anzaldúa's concepts of hybridity, liminality, spirit, spiritual activism, and the recalling of our fragmented parts for deep healing spoke profoundly to me. Similarly, Keating's (2013) notion of post-oppositionality felt simultaneously complicated and resonant.

To explain this resonance, I offer some biographical context. I was born in India and educated in India, Canada, and the U.S. I was aware of my Otherness in Canada, but not until I traveled to the U.S. and attended fully to the pain of being Othered was I able to language it. In the U.S., racialized discourses are located within a black/white binary. Being neither resulted in the erasure, tokenization, or co-optation of my sensibilities during my graduate studies and beyond.

I did not have a single teacher of color in the entirety of my graduate studies, despite completing a graduate certificate in women and gender studies and taking classes in cultural anthropology and the major religions of India. If I voiced a culturally situated perspective in class, it was received with benevolent imperialism (Spivak, 1993). Yet I failed to understand for years that no amount of mastering the texts I was taught would prevent me from being relegated to the margins. Access to the center, if I ever had it, came only via assimilation and tokenization, which required complicity with my own oppression.

This is why Anzaldúa's (1987/1999, 2002) work on occupying liminal spaces with hybrid identities spoke to my hybridized de/colonial sensibilities. Her concept of border crossings meant more than physically crossing boundaries. Her assemblage of personal narratives, hybrid identities, theorization of vulnerable self-excavations, sociocultural discourses, structures of oppression, and connections to worlds beyond the material spoke to my sensibilities as I sought to find my place in academia and beyond, when I was easily and effortlessly Othered based on my accent and ethnicity. Keating (2013) extended Anzaldúa's work by theorizing the notion of post-oppositionality, which cultivated a complicated resonance.

Keating's (2013) work highlights the importance of post-oppositional relationships, or at least working toward them by understanding the corrosive nature of oppositionality. Elsewhere, I state:

> Post-oppositional theorizing then promotes a way to make space for what lies beyond oppositional discourses. In other words, if we are to fight against social injustices in education, how do we do so theoretically, methodologically, and pragmatically, without being co-opted into the binary of the oppressor and the oppressed?
>
> (Bhattacharya, 2016, p. 199)

In reflecting on my own location in oppositional discourses, I identified how we are incentivized to be oppositional in academia and how I have benefited from this incentivization as a scholar who grounds her work in anti-oppression. Yet I have also felt drained and fatigued by this work. Keating's (2013) insights thus became particularly poignant for me:

> Our internal fragmentations – our intra-divisions, as it were – have their source at least partially in the oppositional energies (and the dichotomous thinking behind them) that the groups used to combat social oppression. We can't turn off the negative energies once

we remove ourselves from the battlefield. We take these energies
with us, into our work, our homes, our minds, our bodies, our souls.
They eat away at us, devouring us as we direct this oppositional
thinking at one another and at ourselves. We fragment. We crumble. We deteriorate from within. And then we regroup. We begin
again. And on to the next corrosive battle.

(p. 9)

Post-oppositional thinking could also be read as one that cannibalizes
the very idea it seeks to promote. Yet, Keating does not explicitly ask people to abandon their work or the discourses they find resonant. Instead,
she encourages us to gently push the boundaries of our thinking, discovering what lies beyond the edges of these discourses without dimin-
ishing them, moving with expansive knowledge-making intentions.
De/coloniality and post-oppositionality allow me to engage in interconnectivity in ways that exceed the colonial and opposition-incentivized
practices in academia.

Jia's narrative

I was born in Shanghai, China. As an only child surrounded by strong
women, it rarely occurred to me that I was expected to "behave like a girl"
until I reached school age. Of all the grandchildren, including the boys,
I was my grandparents' favorite; I was the only one they raised until I was
about 10 and moved to live with my parents.

Following college, I taught at various levels, eventually landing a position
in a joint venture company. I came to the U.S. for graduate study in adult
and technical education, a move recommended by my supervisor for career
advancement in human resources. Unfulfilled by this career path, I began
doctoral studies while working with schools in poverty-stricken areas of
West Virginia and refugee students in charter schools in the Atlanta area.

Such close contact with disenfranchised student populations made me
realize that while education can offer pathways to success and foster pros-
perity to address inequity, it also systematically creates hierarchies, deny-
ing many people opportunities for advancement and stripping them of
their self-worth. My teaching, scholarship, and service focus on working
with minoritized groups, in particular girls and young women of color in
education.

I never anticipated that coming to the U.S. would mean being subjected to racialization and other forms of minoritization. These oppressive discourses and materialities were alien to the before-U.S. me. My transnational journey has never ended, despite long ago ceasing any physical travel. A sense of *home* became increasingly elusive as I realized that *back home* exists only in my nostalgia for what existed when I left, while I can never claim full membership in my new *home*.

My accent betrays my American assimilation project, making me immediately identifiable as Other. I remain a foreigner regardless of how long I have lived in the U.S., how much I have paid in taxes, how many U.S. citizens I have educated, or my purchase of a house in the U.S. to better anchor myself. "Go back to your f—— country!" is a racialized attack I experience repeatedly. Always unprovoked, without my uttering a sound, it results simply from my sharing the same space and air with white citizens.

In my home discipline, K-12 educational leadership, I had never encountered any Asian American women until about four years ago. My graduate studies and scholarship led me to recognize the importance of mentorship in fostering leadership development among minoritized people, especially women, and I longed for such a nurturing relationship in my life. Nothing in my master's or doctoral programs at two U.S. institutions addressed Asian Americans' experiences, perspectives, or agency in U.S. society.

In contrast to Kakali, when I attended UGA for my doctorate years later, I was fortunate to have Black women faculty in my home department, in women's studies, and on my dissertation committee. They introduced me to critical theories and research, where I found some validation of my experiences. I eagerly learned about the history of Asian Americans and their historical and contemporary racialization, in addition to reading literature on gender and leadership (Said, 1994; Takaki, 1987; Wei, 2010). However, my qualitative methodology training was entirely with white women scholars.

My dissertation work with Asian American women school leaders used a qualitative multi-case study design. The participants' stories shook me to my core. Generations of living in this country had not made them immune from racism, sexism, nativism, and other forms of discrimination and injustice. Intimately bearing witness to the participants' reliving their experiences and exposing their wounds made me aware of my own wounds. Seeing their passion for educating children, especially marginalized children, furthered my sociopolitical and cultural sensibilities and my interest in the critical framing of social problems and structural oppressions. Meanwhile, I was unaware of the possibilities of my work if I were to pair criticality both in theory and methodology.

Now in my fifth year as junior faculty in a predominantly white institution, I am constantly reminded of my minoritized status by white colleagues and students, often under the cover of coded language such as "cultural relevancy" and "fit" or the facade of well-intendedness, blatant dismissals of my expertise, challenges to my qualifications, isolation, and silencing. I navigate these challenges while hiding the wear and tear on my being. Occupying these in-between spaces of the personal and professional, I continuously reconfigure a sense of "self" and "belonging."

Four mentoring practices

In this section we describe Kakali's mentoring practice and Jia's engagement with that practice. Having worked closely with Kakali, Jia has witnessed these practices as she locates herself within academic, personal, and transnational spaces.

Kakali: shadow work and inner journeying

Shadow work refers to facing the parts of ourselves we repress or fear due to the difficult emotions they might elicit. The phrase refers to Carl Jung's (2014) work on archetypes, in which he identified shadow as a human archetype. Elsewhere (Bhattacharya, 2018), I explain in detail the terrain of my shadow work.

Anzaldúa (2002) extends Jung's notion of shadow by personifying a shadow beast, which she encourages us to encounter to facilitate our collective healing. In my mentoring, I cultivate space for my mentees to engage in inner work in which they excavate their experiences, identify their stuck places, and sit with the difficult emotions that arise. I question whether anyone can offer critical, in-depth, resonant insights from qualitative work without engaging in in-depth investigation into their own experiences and negotiations.

This is not easy work. We must interrogate privileges, oppression, our complicity, the aspirations we chase, and the aspirations we have been afraid to dream about. By cultivating safe spaces and practices of deep listening, a former student concluded that her dissertation should be an autoethnography. A biracial woman, she wanted to unpack the narrative behind her adoption, her pain at her discounted adoption price due to her Black heritage and her birth mother's use of alcohol while pregnant, and her pain in response to her birth mother's decision to keep biological siblings born before and after her. Another student of South Asian heritage identified the pain associated with cultural erasure, something she avoided facing for a long time. She is now tracing the ways in which cultural heritage is being erased amongst second-generation South Asian teenagers.

I do not prescribe how to engage in shadow work, as individuals' journeys and processes vary. However, in assisting mentees, I seek to listen deeply. I have guided people through meditation, contemplative practices such as art-making, labyrinth walking, mindful walking, clay sculpting, and prompted free and creative writing. People chose to do the work that felt safe to them and cultivated their own analytical insights while I partnered with them, holding space and maintaining unconditional positive regard for their process.

Jia: shadow work and inner journeying

The first time I heard the phrase *shadow work*, it sounded distant yet familiar. Although I understood the importance of reflexivity in qualitative research, I considered examining one's own vulnerabilities a requirement of autoethnography, an approach with which I was unfamiliar. My time in the U.S. had taught me to conceal my emotions and vulnerability for self-protection and survival. Opening up even to myself, privately, about deep wounds was not a suggestion I readily entertained, nor did I recognize the relationship between what I carry in my body and being and my scholarship and teaching. I did not know how to do *shadow work* without feeling overwhelmed by pain and self-blame.

Then, at the 2016 International Congress of Qualitative Inquiry conference, Kakali invited me to one of her sessions. I sat in a packed room as she walked us through a shadow work activity using papers and scissors, writing down words, coloring, and drawing things and lines to connect things, all intended for ourselves only. Maybe because I knew it was for myself only and I was surrounded by strangers (who were likely not in my discipline), or maybe because Kakali reminded us that creativity was not a single defined thing and we were all creative in some way, I let go of my resistance and fear, and I participated.

That evening in my hotel room, I reflected. Why is it so hard for me to be open and honest with myself? What am I afraid of? What was different today that allowed me to do so? I wrote two lists: one outlining what I fear, and the other identifying what would happen if I were honest with myself.

The more I reviewed the lists, the more I realized I was never taught to truly love *me*. I was trained to behave, think, even dream in certain ways, and taught the consequences of not living up to such expectations, or of stepping outside the boxes in which I was placed both in China and the U.S. Therefore, I mastered self-censorship. To manage my self-doubt, self-blame, and at times self-hatred, I locked away my wounds, pretending they were not there. I fragmented myself and denied myself the opportunity to heal, connect, and generate the type of compassion and intellectual growth that I admired and yearned.

***

Shadow work and inner journeying represent a juxtaposition of risk-taking and risk-avoiding. With 15–20 students in a graduate-level qualitative research classroom, it is reasonable to assume there is trauma in the room. Therefore, it is sensible to cultivate conditions that call back the parts of the oneself fragmented by trauma. This work is circular, iterative, and recursive, with multiple advances and setbacks. Yet the commitment to journeying creates possibilities for deep insights that guide and align ways of knowing and being and inform inquiry.

Kakali: un/learning established approaches and honoring culturally situated ontoepistemologies

Over the years, I witnessed a proliferation of critical and deconstructive perspectives used to rationalize studies, critique existing literature, and theorize data within qualitative research. However, these approaches are rarely enacted in tangible ways to construct research design, data collection, or, to some extent, data analysis. For example, I have read many theoretically critical or deconstructive dissertations/articles/chapters that use interpretive methods of data collection without addressing the methodological and philosophical incongruities, situating methodology as innocent, acultural, and ahistorical.

What if we reimagined the whole enterprise of inquiry differently? This question becomes especially poignant in working with people seeking to engage in culturally situated inquiry who have not been taught to imagine inquiry beyond privileged qualitative approaches. Much of our training normalizes dominant ways of thinking and engaging in inquiry, to the extent that we cannot detect the incongruencies with our culturally situated ontoepistemologies.

My work here is to bridge culturally situated ontoepistemologies and methodologies. When a mentee wants to explore the influence of informal learning environments on older Black women's sense of spirituality and sexuality, she would not employ focus groups or interviews. Rather, she might approach inquiry as a humble learner seeking lessons from her cultural elders. This cognitive switch allows us to see participants as educators and researchers as learners, altering our relationships and how we author ourselves in those relationships.

Jia: un/learning established approaches and honoring culturally situated ontoepistemologies

As a researcher positioned in critical frameworks, I am no stranger to culturally relevant, culturally situated approaches in educational research. I have yet to fully establish how applying critical and deconstructive methodological lenses can be legitimized in a field like mine, which is heavily focused on practice. Kakali created a series of videos called *What's Hot, What's Not in Qualitative Research?* She invited me to be a guest and asked for my perspectives on methodologies like arts-based research and autoethnography.

I had just started my tenure-track position and taught some qualitative methods courses. I noted my hesitation to experiment with such methodologies, not because I doubted their appropriateness or potential, but because I lacked adequate knowledge and skill to use them well. Now, in my fifth year, I have witnessed many students of color, dissatisfied with restrictive approaches in qualitative inquiry, discover their academic voice through creative, justice-oriented approaches that finally felt culturally congruent to them. When academia remains a hostile space for students of color, anchoring one's dissertation through cultural congruency becomes critical to paving the path for one's academic journey. In this way we model an interconnectivity with a student's cultural understanding and give them the space to trace their connections with their communities as part of their academic work.

***

One way to imagine inquiry differently, to disrupt the white gaze, is to ask, "If I wasn't doing this for academic purposes, how would I really want to learn about X?" Such imagining requires freedom from a disciplinary gaze in qualitative inquiry – always difficult, as most of us have been under some disciplinary gaze throughout our education. Perhaps research design would focus on relationship building instead of extracting data. Data collection might look like cooking with a cultural elder and learning through intergenerational storytelling. To the extent possible, this is what we seek to cultivate: the freedom to identify culturally congruent modes of inquiry and weave ontoepistemologies into the fabric of research.

Kakali: cultivating imagination, play, and whimsy

It is absurd to be forced to normalize interlocked structures of oppression and their manifestation. Confronting such absurdity, we must move beyond an oppressor/oppressed binary and shift our consciousness to other ways of being and knowing. For me, these include accessing absurdity, whimsy, nonsense, and play (Bhattacharya, 2018).

In these spaces I cultivate the imaginal: that which is unthought, unlanguaged, breaks rules of rationality, and brings forth a lightheartedness and joy that illuminate expansive approaches to inquiry. In a previous publication (Bhattacharya, 2018), I highlight the work of Sukumar Ray, a Bengali children's literature author during the British colonization of India. Ray used absurdity, imagery, and satire to inquire into the structure and manifestation of colonizing forms of oppression and Indian citizens' complicities, which, though aiming for survival, cost them their identities, cultural values, and, in some cases, their lives.

At first glance, Ray's work can be read as pure escapism. However, closer investigation reveals the brilliance of a subversive critique framed with nonsense that was unintelligible and untranslatable to the British but could be understood through embodied affect. I argue that nonsense carries:

> no expectation of sense- or meaning-making, rationality, logic, or achieving a philosophical turning point. Nonsense exists for no reason, has no meaning other than that which we ascribe to it. For me, being in nonsense and play generates soul nourishment and sustenance, without any expectation of these ends.
>
> (Bhattacharya, 2018, p. 4)

Soul nourishment is crucial for navigating the colonial academic terrain, with its complicated complicities and entanglements. From this nourished space we can construct different knowledge, engage in spacemaking, become provocateurs, or language fresh insights that resonate with our intended audience. Whimsy, play, and cultivating curiosity and wonder thus became part of my pedagogy, enabling students to experience freedom from rational meaning-making and indulge in simply being.

Jia: cultivating imagination, play, and whimsy

It is no secret that the U.S. education system privileges certain whiteness-centering means of knowledge construction, modes of verbal and written communication, and ways of thinking. Kakali's advocacy of playfulness, whimsy, and nonsense enables us to break the boundaries that imprison our bodies and spirits. Rather than encouraging us to "think outside the box," this approach urges us to think freely in all spaces with our raw sensibilities that are unfiltered and non-widgetized through restrictive academic discursive norms.

In her Introduction to Qualitative Inquiry course, Kakali integrated embodied experiences to explore issues of self, self in relation to others, and justice. She created a Big Ideas video featuring four superhero avatars (Bhattacharya, 2020a, 2020b) with various skills required for qualitative research. The superheroes were visually different from what the west celebrates as heroes. All were non-white, some were gender- and culture-ambiguous, and some were prominently from multiple-marginalized communities. These superheroes conveyed the big ideas for learning qualitative research through whimsy, playfulness, and curiosity. One student, inspired by this playfulness, used *Wizard of Oz* (Fleming et al., 1939) characters to reconfigure her understanding of interpersonal dynamics among professional learning teams in education for her dissertation. The playfulness created a model of possibility for the learner, who extended her creativity to demonstrate complexity through an intelligible metaphor.

Kakali's willingness to engage in play and absurdity inspires me, particularly as I see where she publishes her work. She demonstrates that boundaries are breakable even in highly *rational*, orthodox, and formal spaces, including the most prestigious academic journals. Through witnessing approaches that were absent in my training, I have found anchoring and confidence to do work that breaks my own disciplinary boundaries and methodological restrictions.

***

Doing justice and culturally situated work raises issues that trigger sadness and pain, and we need antidotes to the fatigue that regularly accompanies such work. Joy and absurdity function as antidotes and also as modes of inquiry. Absurdity, whimsy, and disrupting entrapment in the oppressor/oppressed binary allow us to engage in liberatory possibilities we might otherwise forget in the daily bombardment of oppressive forces and consequent materialities.

Kakali: spacemaking for newcomers' raw intelligence and communal knowledge

I strive to make space for the raw intelligence and communal knowledge that newcomers, especially those from multiply marginalized communities, bring to academia. Raw intelligence, here, refers to the knowledge students bring in as they enter academia, which often get stifled, erased, or transmuted through Westernized normative academic discourses. Academia is a colonizing enterprise (Smith, 1999/2002), with practices that disenfranchise multiply marginalized communities. These communities' raw intelligence and knowledge are dismissed as either backwards, lacking academic value, or cleverly co-opted, repackaged into obscure language, and situated behind paywalls, reinforcing exploitation and marginalization.

Newcomers must navigate between the voice and sensibilities with which they informed their lives prior to entering academia and the expectations of scholarly practices. During this process, impostor syndrome (Studdard, 2002) and various types of battle fatigue (Chancellor, 2019; Corbin et al., 2018) arise. As an educator and mentor, I have a responsibility to not widgetize students to perform to academic norms while sacrificing their voice and identity. To the extent possible, I try to incorporate students' embodied knowledge into their work. The assumption that only academics theorize is incorrect. People outside academia theorize, but that work is languaged differently, especially when they theorize from the street (Bhattacharya, 2019). Within qualitative research, there have been caustic battles between those who use certain high theories and those who do not. Theorizing from the street refers to people living their lives on the street or outside of academia navigating multiple structures of oppression daily, who have theorization skills that should not be hierarchically understood in relation to discourses of high theories in qualitative research.

A de/colonizing approach to qualitative research and mentoring requires tracing how colonizing sensibilities are embedded in academia and how I might disrupt them in collaboration with newcomers. We may cite differently, or create conceptual maps that integrate communal knowledge as a framing device, or engage in ways of knowing and being that are not privileged in academia. We may offer different language and approaches, or bring into academia the sensibilities that ground us in our everyday spaces.

Jia: spacemaking for newcomers' raw intelligence and communal knowledge

Carving out space through mentoring for the raw intelligence and communal knowledge of multiply minoritized people in predominantly white spaces requires taking professional risks, especially when mentors are also multiply minoritized. Many of Kakali's doctoral advisees are from my home discipline, K-12 educational leadership – a field that remains predominantly Eurocentric and masculinist in its theories and practices. Most faculty are white, and exposure to *non-mainstream* curricula is limited to a few designated diversity and ethics courses.

From Kakali's work with minoritized students, I have learned her ultimate respect for their indigenous and culturally situated ways of being and thinking and an excitement for integrating such cultural rootedness. Kakali often urges advisees doing culturally situated work to "cite your own people." This potent advice recognizes the power difference between students and faculty; while functionally it may give permission, it also transparently acts to center and validate erased or minoritized perspectives in academia.

Sometimes people say, "You don't know what you don't know." In graduate school, I did not know about Asian American feminism or transnational feminism, which I could have leaned on to understand my experiences and identities. My advisor and committee were supportive; however, they were not culturally situated to direct me to such literature. In retrospect, I wonder how my graduate school experience and dissertation work with Asian American women school leaders would have differed had I been grounded in an ontoepistemology that was more aligned with my background and identity.

***

Knowledge making should not occupy a privileged space, open only to those who obey the widgetized demands of academia. Granted, this form of mentoring is not for everyone; some people align well with the status quo, and others are simply not disruptive types. However, those who experience academic expectations as incongruent with their ontoepistemologies may welcome academic pathmaking that is commensurate with their sensibilities.

Mentoring as a radical act of interconnectivity and love

Academia can be isolating, particularly for those navigating multiple structures of oppression. The practices of mentoring shared here foster radical interconnectivity and draw on what bell hooks (2000) calls an ethic of love. hooks (2000) encouraged us to awaken to love by releasing our obsession with power and dominance and believing the heart as well as the mind occupies a seat of learning.

Anzaldúa (1999) cautions that allowing lovelessness to guide our lives leaves us feeling isolated and alone. For hooks (2000), love is not a mystery; embodying love requires simply cultivating an awareness of love. Anzaldúa and hooks advocate facing our wounds and engaging in shadow work and inner journeying to cultivate an awareness of love and, by extension, interconnectivity. In cultivating an ethic of love, one moves toward others who are similarly situated, mitigating feelings of isolation. This becomes a radical move to make in academia, when we are incentivized by oppositionality to carve our spaces.

For mentors, cultivating an awareness of love means showing up for each other and our students, offering them our full presence, and telling the truth even when that requires a difficult conversation. It means building the capacity for difficult conversations without blame or judgment, detaching from the outcome while accepting the uncertain consequences of these discussions. Mentoring from these sensibilities requires creating a sense of belonging, sharing love, sifting through multiple truths, embodying compassion, dreaming of colonization-free futures, resisting injustices, and incorporating regular periods of rest and retreat; so that, even if we must sometimes bend, we ensure that we will not break.

To engage in acts of loving awareness, we must interrogate our own stances and journey within. We must understand how suffering shows up in our bodies, hearts, minds, and spirits. Where do we hold suffering in our being? Where do we resist it? Where have we silenced it? Through such mindfulness we offer ourselves much-needed love and compassion, ena-bling us to do the same for our mentees, which is disruptive to imperialistic forces.

Sometimes cultivating an awareness of love requires sitting with our mentees in stillness and silence, slowing our cognitive and affective pro-cesses to simply be in each other's presence before illuminating possible paths. In sitting with stillness, we connect in ways that exceed language, understanding one another's loving presence in intuitive, embodied ways, as if creating a cellular resonance. When we fail or make mistakes, we humbly accept responsibility, which is another way of creating loving awareness. This prevents others from feeling victimized by our actions and allows for grace, which is not always a priority when scapegoating offers the potential for individual academic rewards. These are acts that exceed oppositional-ity, move beyond imperialistic benevolence, and offer possibilities to build connectivity.

Radical interconnectivity requires disrupting certain fundamentally col-onizing and oppositional discursive practices in higher education. Addi-tionally, understanding mentoring as radical interconnectivity and love requires a deep state of gratitude for our journey. Mentoring in academia, especially in doctoral education, allows us to bear witness to a fellow human being's journey of earning the highest degree on the planet. We have front-row seats as people achieve major milestones in their lives. That, above all, is a gift we must cherish.

Within the context of qualitative inquiry, mentoring from the sites and bodies that we occupy require us to unlearn some of the whiteness-centering discursive practice and disrupt that we have normalized and internalized. We have documented Kakali's mentoring practices and the enactment of those practices as witnessed/experienced by Jia. Jia's narratives directly address the influence of the practices on her orientation to qualitative inquiry and her witnessing of students' qualitative research projects after being at the receiving end of the mentoring practices. The practices become heuristic for thoughtful cultivation of depth, criticality, positionality, resist-ance to oppressive discourses and materialities, and spacemaking for the freshness of knowledge newcomers bring to the field without committing violence and erasure to their culturally situated sensibilities. The practices help in understanding ontoepistemic orientations, articulating justification of theoretical, methodological frameworks, analytical processes, navigation of ethical crossroads, and the decisions made for representing data in published places for qualitative inquiry.

In our informal learner-learned relationship, we engaged each other through a practice- and process-oriented approach, without any specific predetermined outcome. But at some point, we recognized that what Kakali un/learned in her journey of being a qualitative methodologist of color is what Jia would need to learn both in her research and teaching qualitative research classes to educational leaders, even if she did not fully identify as a methodologist. We fell into a culturally familiar structure of friendship, community building, and acknowledgment of experience and wisdom, with the desire to understand and unpack lessons learned from paths traveled. Indeed, there was a power difference based on Kakali's rank and years in academia, but that difference was leveraged to create safe and supportive spaces for Jia.

On the final day of Kakali's classes, she invites students to choose a polished stone from her collection. Asking each student to hold the stone with their eyes closed, Kakali guides them with a set of affirmations, visualizing a future in which students are purposeful, agentic, and transformative through their work. She walks them through the stuck places that may arise and assures them that drawing on their interconnected network of peers, mentors, ancestors, muses, and anyone else as they see fit will provide what they need to navigate barriers. Students keep their *research rock* with them as they write and defend their prospectus and dissertation, present conference papers, respond to journal reviewers, or engage in any other act for which the research rock can offer strength. An inanimate object, embedded with an awareness of love, becomes the medium through which past, present, and future culminate in building various connectivities.

The four mentoring practices embody these radical acts of interconnectivity and love through which we build relationships as qualitative research educators and guides with those who come to learn with us, with whom we learn, resist injustices, do inner work, journey together, honor the knowledge we bring to higher education while sharpening our craft, unapologetically center and integrate culturally situated ontoepistemologies, theories, and methodologies, build community, center humility and gratitude instead of entitlement, and cultivate joy, playfulness, and whimsy. Ultimately these practices evoke the question: What are we leaving behind for the next generation of qualitative researchers to learn?

References

Anzaldúa, G. E. (1999). *Borderlands la frontera: The new mestiza* (2nd ed.). Aunt Lute Books. (Original work published 1987)

Anzaldúa, G. E. (2002). Now let us shift . . . the path of conocimiento . . . inner work public acts. In G. E. Anzaldúa & A. Keating (Eds.), *This bridge we call home: Radical visions for transformation* (pp. 540–577). Routledge.

Bhattacharya, K. (2006). *De/colonizing methodologies and performance ethnography: Extending and legitimizing qualitative approaches to "scientific" inquiry*. Paper presentation, American Educational Researchers Association, San Francisco, CA.

Bhattacharya, K. (2009). Othering research, researching the other: De/colonizing approaches to qualitative inquiry. In J. Smart (Ed.), *Higher education: Handbook of theory and research* (Vol. XXIV, pp. 105–150). Springer.

Bhattacharya, K. (2015). Diving deep into oppositional beliefs: Healing the wounded transnational, de/colonizing warrior within. *Cultural Studies <=> Critical Methodologies, 15*(6), 492–500. https://doi.org/10.1177/1532708615614019

Bhattacharya, K. (2016). Dropping my anchor here: A post-oppositional approach to social justice work in education. *Critical Questions in Education, 7*(3), 197–214.

Bhattacharya, K. (2018). Walking through the dark forest: Embodied literacies for shadow work in higher education and beyond. *The Journal of Black Sexuality and Relationships, 4*(1), 105–124. https://doi.org/10.1353/BSR.2017.0023

Bhattacharya, K. (2019). Theorizing from the street: De/colonizing, contemplative, and creative approaches and consideration of quality in arts-based qualitative research. In N. K. Denzin & M. Giardina (Eds.), *Qualitative inquiry at a crossroads: Political, performative, and methodological reflections* (pp. 109–126). Taylor and Francis.

Bhattacharya, K. (2020a). Nonsense, play, and liminality: Putting postintentionality in dialogue with de/colonizing ontoepistemologies. *Qualitative Inquiry, 26*(5), 522–526. https://doi.org/10.1177/1077800418819624

Bhattacharya, K. (2020b). Superheroes and superpowers: Contemplative, creative, and de/colonial approaches to teaching qualitative research. In P. Leavy (Ed.), *The Oxford handbook of qualitative research* (2nd ed., pp. 1163–1182). Oxford University Press.

Chancellor, R. L. (2019). Racial battle fatigue: The unspoken burden of Black women faculty in LIS. *Journal of Education for Library & Information Science, 60*(3), 182–189.

Corbin, N. A., Smith, W. A., & Garcia, J. R. (2018). Trapped between justified anger and being the strong Black woman: Black college women coping with racial battle fatigue at historically and predominantly White institutions. *International Journal of Qualitative Studies in Education, 31*(7), 626–643. https://doi.org/10.1080/09518398.2018.1468045

Denzin, N., Lincoln, Y., & Smith, L. T. (Eds.). (2008). *Handbook of critical and indigenous methodologies*. Sage.

Fleming, V., Cukor, G., LeRoy, M., Taurog, N., & Thorpe, R. (1939). *The Wizard of Oz* (M. LeRoy) [Film]. Metro-Goldwyn-Mayer.

hooks, b. (2000). *All about love: New visions*. Harper Collins.

Jung, C. G. (2014). *Collected works of C.G. Jung, volume 9, part 1: Archetypes and the collective unconscious* (G. Adler & R. F. C. Hull, Trans., 2nd ed.). Princeton University Press.

Keating, A. (2013). *Transformation now! Toward a post-oppositional politics of change*. University of Illinois Press.

Rhee, J., & Subreenduth, S. (2006). De/colonizing education: Examining transnational localities. *International Journal of Qualitative Studies in Education, 19*(5), 545–548. https://doi.org/10.1080/09518390600886189

Said, E. (1994). *Orientalism*. Vintage Books.

Smith, L. T. (2012). *Decolonizing methodologies: Research and indigenous peoples* (2nd ed.). Zed Books. (Original work published 1999)

Spivak, G. C. (1993). Can the subaltern speak? In P. Williams & L. Chrisman (Eds.), *Colonial discourse and postcolonial theory* (pp. 66–111). Columbia University Press.

Studdard, E. S. S. (2002). *Behind the mask of success and excellence: Impostorism and women doctoral students* (Doctoral dissertation). https://athenaeum.libs.uga.edu/handle/10724/29439

Takaki, R. T. (1987). *From different shores: Perspectives on race and ethnicity in America.* Oxford University Press.

Wei, W. (2010). *The Asian American movement.* Temple University Press.

A MENTORING MOMENT WITH MY MENTOR

Ying Wang

More than a decade ago, my mentor wrote about her frustrating learning experiences as a Peace Corp volunteer in Papua New Guinea. As she put it, "I observed and participated in these daily activities without recognizing them as learning. This non-explicit, nondirective way of teaching did not fit my notions of education" (Huckaby, 2004, p. 79). Ironically, despite her frustration about Papua New Guinean non-explicit teaching, my learning how to be an ethical qualitative researcher mostly came through anecdotal mentoring moments that echo non-explicit ways of teaching and that would not be necessarily considered as teaching in our inquiry seminars' syllabi. Due to the space limit, I will share one such salient moment I have had with her.

In spring 2016, my mentor asked me and another fellow graduate student to travel with her and interview some teachers and parents in another state. Because of the person she knew from one United Opted Out conference, we got acquaintances to people and easy access to important meetings in the local context. However, the person, though being an activist herself, was not willing to be filmed or interviewed. Back then, I did not know the person's nuanced stance and asked my mentor at one point, "Shall we interview such-and-such as well?"

"I don't think so. I have invited her twice in our earlier conversations, and she kinda declined to speak in front of the camera. I mean, if I push further, we may get the interview. But that's it." She replied as she was packing filming equipment for the day's events.

I remembered she said that in a very stern voice and with seriousness on her face. I gathered "that's it" means coercing the person on board would sabotage the mutual trust and relationship already established between my mentor and the activist; meanwhile, "pushing further" is unethical, even though it would not reflect in documentation or in the final write-up. The potential participant's concerns and vulnerabilities could have been brushed aside or persuaded, had my mentor pushed further. However, for my mentor, personal concerns are unconditionally more important than whatsoever stunning interviews that can be exploited for the research purpose. Coherent with her understanding power as dynamic, my mentor renders power as "not a possession, but an ability with an understanding that all beings are within relations of vulnerability and power" (Huckaby, 2013, p. 578). Also, it seems to her that how to position oneself in field research *per se* is a site worth epistemic inquiries. She wrote much earlier than the aforementioned trip:

> As an academic and theorist, I have made and continue to make use of Foucault as I wrestle to understand and engage relations of power. However, the intimate curriculum of my pre-extra-academic life, like that of most people, is at work and mediates my interpretations and use of Foucault and now fuels my challenges to power as I am re-theorizing it through an attention to vulnerabilities.
>
> (Huckaby, 2013, p. 569)

As a Foucauldian scholar, she firmly believes power is situational and strategic; power is distributed discursively in the field. As such, she holds that power comes along only when vulnerabilities are shared and fully acknowledged. She humbles herself as a human being when in the field; she is always ready to help wholeheartedly and learn genuinely from others. That field trip with her became instrumental to my understanding of what power dynamics could look like during research and what it means to be an ethical qualitative researcher beyond an approved IRB protocol.

References

Huckaby, M. F. (2004). When worlds collide: A postcolonial and critical view of Western education in Papua New Guinean village schools. *The Journal of Curriculum Theorizing, 20*(4), 75–90.

Huckaby, M. F. (2013). Much more than power: The pedagogy of promiscuous black feminism. *International Journal of Qualitative Studies in Education, 26*(5), 567–579.

Mentorship as Radical Care

Elliott Kuecker

We don't always treat each other well in academia. There is an indoctrinated sense of austerity of time, ideas, faculty job positions, publication capacity in a journal, leadership roles. It is no wonder that graduate students regularly experience "days of crisis," as a friend described to me, in which terrible interactions from known or unknown academicians trigger the need to abandon academia. This friend shared a meme, which read "Blind Peer Review," that showed a kid blindfolded, vigorously pounding away at a dangling pinata.

In a space like this, the most likely form of care one receives is what Chris Ingraham (2020) has called "gestures of concern": "an expression into form of an affective relation" (p. 1). These rhetorical moves help contribute to a *sense* of care, but amount to a Get Well Soon card, in that they contain no actions. Academic interpersonal relationships are often more affected toward "[t]hat anxious posturing, the vigilant search for mistakes and limitations, the hostility that crushes the hesitant new idea, the way that critique becomes a reflex" (Montgomery & Bergman, 2017, p. 20).

I see my mentor as radically interrupting these norms, revealing a kind of care that consciously rejects hoarding time and ideas. She surprised me when she replied to an email I wrote her at the end of a semester, asking about a small archive from an experimental elementary school in her

 DOI: 10.4324/9781003022558-17

town of Bennington, Vermont. She said she knew the school and had once intended to do research there.

We corresponded about these archives and soon we were meeting twice a month, online, to look at the digitized drawings children had created at this school, and talking about our own children's art. While the opportunity to research among the actual children has passed, given the school's closure, we found ourselves making sense of the traces left, reckoning with the temporality of childhood, while working through what a qualitative approach to historiography might be. We collectively filled over 50 pages of notes from readings on art history, archival representation, and materiality in art.

Let me know when it is too much, I begged her. But summer turned into fall and we were still at it, writing an article together on how qualitative archival research is an act of renewal. Her decisions to meet with me were affirmations of care for our project rather than her own – articles, grandchildren, home improvement, reading. As a senior scholar and involved family member, any of these might have generated more legible accomplishments for her. In the innumerable decisions about how to spend her time each week, she continued to choose to gather around this project with me. Such a thing is nothing short of a radical ethics of care.

References

Ingraham, C. (2020). *Gestures of concern*. Duke University Press.

Montgomery, N., & Bergman, C. (2017). *Joyful militancy: Building thriving resistance in toxic times*. AK Press.

7

The importance of relationships in mentorship and methodological identities

Stephanie Anne Shelton, April M. Jones, Kelsey H. Guy, and Boden Robertson

None of the four of us would have ever imagined that we would be here, now, together. As three of us pursue PhDs in qualitative research and one seeks tenure as an assistant professor of qualitative research, we find ourselves a bit baffled as to how we ended up here. Three of us are first-generation college students, and all four of us began our academic journeys in departments and colleges outside education and research methodologies. No one would have convinced any one of us, only a few years ago, that methodology would be our trajectory, and perhaps even more impossibly, that being here would be *right* for us. As we considered how we arrived at where we are, we agreed on a single-word explanation: relationships. Our personal connections directly shape and inform our professional ones, joining us together in powerful and productive ways as colleagues (#qualleagues), mentors, and mentees. Additionally, explicitly exploring the ways that personal connections inform mentorship humanize academia in valuable ways and demonstrate the importance of forming and celebrating relationships in higher education such as what we describe here.

Limited discussions on mentoring and relationships in qualitative research

Despite ever-growing scholarship in the field of qualitative research, there are surprisingly few discussions on mentoring and relationships and how

 DOI: 10.4324/9781003022558-18

they matter to individuals and to the field. As we worked to understand how our own experiences fit into the larger body of qualitative inquiry, we were surprised to find that most literature specifically examining these topics did so within organizational contexts. For example, many scholars employed qualitative methods to understand mentorship within specific institutions or settings (e.g., Coppin & Fisher, 2016; Gunuc, 2015). Others explored mentorship relative to designing educational materials (Çobanoglu et al., 2017; Dönütü & Uygulamalar, 2020; Sassi & Thomas, 2012). A few considered the ways that these factors shaped students' engagement with specific methods, such as interviews (Hoskins & White, 2013) and performative narrative (Chawla & Rawlins, 2004).

Unquestionably, these are important discussions, but they do not fully engage with how *necessary* relationships and mentorship are to qualitative inquiry and, more specifically, to qualitative researchers – novice and experienced. As we four sat around a table at a coffee shop, we reflected on how our personal connections had mattered, and what they had done to/for us. Prior to this meeting we had, as perhaps is the norm of nerdy academics, created a Google doc in which we jotted down thoughts relative to our relationships with one another and qualitative research. We found our multiple experiences and perspectives constantly folding into one another, overlapping in surprising but meaningful ways. Each time that one of us reached back to how we had found qual, we immediately and simultaneously found ourselves noting the ways that the others in our group had helped to shape each journey.

Relationships and a pedagogy of tenderness

As we considered ways to write together and honor our intertwining paths, relationships, and experiences, such an enterprise seemed initially impossible. How would we weave together four separate/enmeshed/messy trajectories and emphasize relationships as the core of those experiences? We found discussions on a "pedagogy of tenderness" meaningful in these efforts. To clarify, this "tenderness" is not a concept that involves "cooing" and cartoon hearts (Thompson, 2017, p. 1). Instead, this approach was born out of synthesizing elements of feminist, queer, and women of color frameworks – all frameworks that matter to us four individually and collectively – to disrupt norms, de-center authority, and embrace the personal as valuable to educational and professional interactions (Schulz et al., 2011; Thompson, 2017).

A pedagogy of tenderness emphasizes the ways that relationships matter in educational experiences and offers:

> an embodied way of being that allows us to listen deeply to each other, to consider perspectives that we might have thought outside our own worldviews, to practice a patience and attentiveness that allow people to do their best work, to go beyond the given, the expected, the status quo. Tenderness makes room for emotion.
>
> (Thompson, 2017, p. 1)

This approach within education emphasizes that "significant learning will not take place without a significant relationship" (Brown, 2008, p. 10) between those involved. So many academics "have been disciplined not to include . . . emotions in our stories and analysis" (Thompson, 2017, p. 2), and thus meaningful relationships with higher education peers often do not factor into the ways that we strive to learn or mentor. However, leaving these relationships, which are necessarily tied to our emotions and personal selves, out of our academic lives dissects our "whole selves" and carves academia out to be separate from our "physical, spiritual, and intellectual development and well-being" (Schulz et al., 2011, p. 56).

Choosing to acknowledge, celebrate, and tap into these connections between one another helps to construct "a community of learners" that disrupts "faculty as the ultimate and only experts" (Thompson, 2017, p. 3); instead, a pedagogy of tenderness emphasizes learning and mentoring based on "lived experiences in the world," while encouraging conversations and connections that may be spontaneous or planned but always work to be honest and open (p. 6). We draw on these tenets to explore our relationships and experiences with mentorship within qualitative research. Our connections offer "bigger ways of understanding ourselves" (p. 6) as individuals and collectively. In order to actively engage in this approach, we collaboratively wrote this chapter, from the beginning until now. We shared documents, actively commenting on one another's thoughts/questions/contributions, adding our own views and experiences, so that we and our writing were in constant conversation with one another – whether actually vocally or through typing. Additionally, to emphasize the importance of the personal and everyday within a pedagogy of tenderness, our collaborative work emphasized narratives: stories of ways that we had found, shaped, and mentored one another. These stories do not follow a specific linearity but instead seek to emphasize connections that shaped our engagements with one another and with qualitative research.

Falling into Qual

Stephanie, about 15 years ago: I am honored to serve on the State Superintendent's Teacher Advisory Board. Seated around the room are a smattering of K-12 teachers like me, and state legislators. "We can finally reshape education to serve kids," I think. The meeting begins with a proposed bill changing high school graduation requirements, and it is immediately clear to *every* educator that this proposal will disproportionately affect low-income students. Teacher after teacher shares an example of a child who would be crushed under this legislation if it becomes law. After several minutes, one of the legislators clears his throat and derisively tells us, "These are *stories*. We need *data*." I feel my heart race and my stomach clench; stories *are* data, damn it. But I don't feel able to articulate *why* they are.

Stephanie, a few years later, beginning a PhD program: I stare at my registration screen, absolutely clueless. A newly enrolled PhD student, I've called my advisor for help. "How in the hell do I know what research classes I'm supposed to take?" I ask. He responds, "Well, what do you want from research?" I think back to roughly five years ago and tell that story. "I want to know how to make idiots like that guy understand that stories *are* data," I finish. "Okay, qualitative research it is, then." I have no idea what this "qualitative research" is, but I'm excited.

Kelsey, fall 2012: "Wait, wait, wait – they did what with the outliers?"

I'm sitting in my ethology (animal behavior) course, a biology major nearing graduation. My professor, my favorite instructor and someone I unflinchingly respect, looks up; I rarely ask questions in this class. The study we've been discussing has a section where the authors explain that some of the outlying data had been removed from the results. We were just about to move on to the next study – am I the only one bothered by this segment?

"I just don't understand how they can throw those results out; isn't that tampering with the data? Shouldn't those outliers be explored rather than tossed?"

"Not when the authors are trying to prove that their method is novel and has significant results," my professor replies.

"But if the outliers were included," I press, checking the table of figures to make sure I'm right, "their results wouldn't be significant."

"Then that's probably why they analyzed the findings differently than you're proposing," my professor says. He can tell I'm dissatisfied with the answer, but class is almost over. He gives me a "What're ya gonna do?" kind of sad smile, and we move on.

Boden, fall 2017: Enrolled in a PhD program in second language acquisition (SLA), I have little direction for my coursework due to overburdened advisors. So, I register for two classes: critical theory and qualitative methods. At first, I don't get qual. I mean, our professor asks us to interview someone and then draw or paint something about the experience. I'm like, "Wait, *this* is what doctoral coursework is? Shouldn't I be learning something instead?" Little do I know, I am.

Kelsey, fall 2017: "So I have to take qual? I can't just take quant?"

"No, you have to take two courses of each."

"Ugh, this is going to be the worst," I think as I shift on my advisor's stiff couch. "Okay then, let's go ahead and do those next semester. I'd rather get it over with sooner rather than later."

"Boden is taking Qual I this semester," my advisor offers. "You should talk to him and see what he thinks; I hear he's enjoying it."

Kelsey, spring 2018: I walk into my Qual I classroom skeptical and stubborn, but a little bolstered by Boden's encouragement. He described it as "challenging but extremely useful" – that every bit of class time had a purpose. "Whatever," I think, mentally rolling my eyes, "this is going to be an absolute mess."

And it is. But it is a glorious mess.

Boden, spring 2019: This course combination of SLA and Qual I has been lethal to my intended program of study. Giving critical theory *and* qualitative methodology to someone already suspicious of tradition? I have begun questioning well-established methodologies in SLA. I see scholars recycling methodology, re-producing pre-existing knowledge, and therefore accomplishing little. I want to be part of unlocking something new and noteworthy. I am no longer interested in the same-old-same-old.

So, I dig into the literature, intending to make a case for the importance of qualitative methodology in SLA. Presenting on this argument at a conference, I notice that the keynote speaker is in the audience. His philosophy: quant or bust. During the Q&A session, he attacks me with a flurry of positivist buzzword soup: "generalizability," "reliability," "validity." I attempt a standard defense, that these concepts are not qualitative research goals; rather, its *strengths* lie in these presumed "shortcomings." He wants nothing of it: "The replication of results is the cornerstone of the scientific method!" he exclaims. I want, tone dripping with attitude, to respond, "According to whom? *You?* Who decides on these 'cornerstones'?" The pressure of maintaining decorum helps me hold my tongue. In my silence, he adds something about qualitative research being untrustable and about the flaws of the intrusive researcher's subjectivity. My researcher mind rages, "Trust? TRUST?!? You trust these numbers, you trust arbitrary confidence intervals, you trust the casting aside of so-called outliers? I want to know these outliers. I want to know their stories. But for some reason *I* have to convince *you* that *I'm* trustable." Instead, I swallow my pride and smile, "Thank you for that question. It's something I'll take into consideration." He smiles back, pleased with himself and his effort to break the will of those challenging traditional approaches.

Boden and Kelsey, later the same day: Kelsey and I attend the keynote's address. He presents all of these numbers. Despite both of us admittedly knowing just enough about quant to be dangerous, we immediately see flaws in his analysis. Then the bomb drops: "I have to thank the quant research lab at my university for these numbers, because I don't do any of the analyses myself," he states. My jaw drops. I nearly scream again, "TRUST?!? TRUST?!?"

Kelsey and I have a long conversation about my program and my leanings toward qualitative methodology. I feel lost. Kelsey says something that will be life-altering, "There's an available assistantship in the qual program." My future begins to look very different.

April, spring 2017: I had been employed full-time at my agency for over ten years and working in my current position for more than five years. I cannot pinpoint what has prompted/initiated the internal desire for a shift in my normal day-to-day; maybe it's boredom. But after three years, I yearn for a new experience and intellectual challenges. My solution: I'll take a doctoral-level research course. This decision doesn't materialize out of thin air. I have always had a passion for research, but it has been seven years since I've actually engaged in tangible research endeavors; you know, data collection, analysis, report writing. I'm confident that my brain needs a serious tune-up before I engage in any quantitatively based research projects.

Which only prompts more questions: What course should I take first? What course would be a fairly painless transition back into higher education? I have already earned a master's degree in social work at the very institution where I plan to take courses, so will it be weird if I enroll in a master's-level course at my age? How about doctoral-level courses? Can I enroll in those courses without being a doctoral student? So many musings and no one to ask. So, unsurprisingly, I do some research.

For two days, I review information on the university's website before ascertaining that, as a non-degree-seeking student, I can enroll in doctoral-level courses and avoid master's-level classes. "Whew, what a relief!" Now the only questions that remain are: What course? And, what won't give me a headache or make me feel totally incompetent? More research. Another two days and I narrow my list to three quant courses in social work. And yes, two days – because none of the course descriptions excite me. But I have to select *something*.

For a week I carry around my notebook listing my top three course choices. With my list, I solicit a couple of colleagues' opinions, one who has earned a PhD and one who is a doctoral student on assistantship with our agency. My PhD colleague points out that I do not have to limit myself to social work courses, that there are other quant classes in other colleges. However, my discussion with the graduate assistant totally shifts my plans and, ultimately, my trajectory. He describes his recent experiences in a qualitative research course in the College of Education and recommends that I look at some of those courses. As I do so, I realize why *none* of the quantitative courses intrigued me: I am not a quantitative researcher at heart, and I have only engaged in quant out of training and practice. Honestly, I had not realized that there are other ways to research. That night, for hours, I review qual course descriptions and finally, after resisting the urge to fall back into the familiar with quant, enroll in Qualitative Research I.

April, fall 2017: It is the first course meeting, and I am nervous. Suspecting that I will be the only person in class not actually enrolled in a doctoral program, the initial introductions prove me right and cause a small jab of insecurity. I push that feeling aside and remind myself that I work full-time and have made the decision to continue my education. I am here now, and I am capable . . . "*I think?*" It is the professor's introduction to the course and initial class discussion that settle my nerves. I absorb little of what is said, but now I'm tingling with excitement. Qualitative research is my jam!

It meshes well with my natural inclination to always ask "Why?" and it provides me with new means to seek understanding.

April, later that semester: This is a challenging semester. Despite a vaccine, I have had the flu; my work tasks have increased; my brain is recalibrating to being a student again. But it is wonderful! So wonderful, that I think, "I should pursue the qualitative research certificate. It'll legitimize my new-found research interest, and advance the research projects that energize me." I attend an information session to learn more about the qual certificate and the opportunities it offers. There, I talk with other students, all of whom are enrolled in a PhD program. "I am not ready for a doctoral program," I think. After all, my brain and my personal life are just re-orienting to course work. Enrolling in a PhD program with more than one class at time is out of the question – *"Right?"*

As the semester moves forward, I encounter others' questions about my earning a doctorate. Many assume that I am already enrolled in a program, but I never linger on the idea of actually pursuing a PhD for more than a day.

April, spring 2018: I am enrolled in my second qualitative research course: Case study. It is evident to others and myself that I am hooked on qual, but where should I go from here? The qual certificate? Or . . . PhD? Definitely the certificate: it will allow me to work full-time and accomplish a goal simultaneously. However, my fondness for the qual faculty and the pointed comments from Stephanie, my professor, disrupt my certainty. The feedback on my assignments includes comments like, "You are a natural methodologist!" or "You're totally made for doctoral work, whether you believe it or not." I struggle to believe it. Why a PhD? Do I want to be a professor in higher education with extensive research demands and pressures to publish? What opportunities exist outside the academy for qual researchers?

Once again, in true qualitative fashion, I question others and myself. Using a class assignment, I conduct a focus group of doctoral students to explore my concerns. I ask them the questions racing around in my head, and their responses help me to check off various items on my pros-and-cons list. The qual certificate and qual PhD program are tied.

April, fall 2018: I am enrolled in my third qualitative research class. Yes, my third course, and *no,* I have not made a decision. Armed with answers to my previous queries, I find myself asking another question – probably one of the most important questions: If I apply for the PhD program and get accepted, will I be able to work *and* enroll full-time?

April and Kelsey, fall 2018: Kelsey and I are both enrolled in Qual II and employed full-time. The difference is that Kelsey is also a full-time student, as I meander along with my one-course-at-a-time strategy. I need to understand how she is managing it all – the work, school, and personal and social life. Is it possible to juggle all of those areas and maintain a modicum – or even just semblance – of sanity? I know others who've earned PhDs while balancing all of these elements; the experience has gone poorly for nearly all, and I am apprehensive and confused.

Perhaps a natural ethnographer, I observe and listen as much as possible. In doing so, I learn more about Kelsey. I like her: She is smart and knowledgeable, and I value her perspectives. During an in-class activity, Kelsey helps me to process a concept proffered by Deleuze; it takes about ten minutes, but it finally clicks. I think, "If she can help me to tackle Deleuze, then I feel comfortable asking Kelsey questions about getting a PhD. Her responses would help to answer my lingering questions." She cannot adequately answer these questions during brief class breaks, so I ask if I might essentially pry into her personal life outside class. After her generous "yes," we plan to meet at a coffee shop next week.

April and Kelsey, a week later: This informal meeting with Kelsey is pivotal. I am amazed by her ability to juggle it all, and I am oddly comforted by her honesty that all is not perfect. What becomes evident as we talk is that I want to earn my PhD. How in the world that will play out is unclear, but I know that the qual certificate is no longer enough. After this meeting, I schedule an appointment with Stephanie to determine the next steps for applying to the qual PhD program.

April and Stephanie, the same week: In true form, I ask more questions – this time to Stephanie. To my pleasant surprise, I do not seem to annoy her, a fact that makes me want to be a part of the qual community even more. For over 40 minutes, we consider each of my questions. I have at least ten questions in my notebook, and I realize that I was not expecting this level of comfort with her. She reminds me that this is a big decision and that my questioning is a "completely reasonable response." Like me, Stephanie worked full-time for several years before enrolling in a doctoral program, so she understands my nerves and slight apprehension. I begin to feel at ease, and my brain insists, "Take the next step, April." But I have a remaining question: "Are any of the qual staff planning to leave anytime soon?" These faculty are a major part of my decision-making, and I do not want to enroll and then find them all gone. Stephanie smiles and replies, "No, no one currently plans to leave." I begin to recognize my connections with Kelsey and Stephanie as something more than sage conversation or advice; these interactions are mentorship – through wonderful, emotional, transformative relationships.

Reflecting on our connections

As we began to collectively weave our multiple experiences with qualitative research into a single narrative, we were struck by how our experiences accentuated the degree to which mentoring – at its best – is relational and non-hierarchical. By definition, because mentoring is generally a more-knowledgeable person advising a less-knowledgeable person, mentoring is presumably inextricably intertwined with power structures. But many of the most valuable conversations that we recalled were peer-to-peer: Kelsey begrudgingly enrolling in qualitative research, only because Boden reassured her; Boden moving into the qual PhD because Kelsey supported and encouraged him; April considering a PhD at all only because Kelsey and Stephanie readily shared their own experiences. This disruption to power is an important element of a pedagogy of tenderness. To be clear, there is no romantic pretense of fully decentering authority. Stephanie is a professor, and April, Kelsey, and Boden are doctoral students. What a pedagogy of tenderness *does* work to do, though, is "grapple with privilege" and recognize locations of power (Thompson, 2017, p. 4); through that recognition, we could use that authority when it was helpful and reject it when it interfered with relationship-building and empowerment. So, for example, we needed Stephanie's faculty status to work through program admissions funding details, but the interactions still worked to emphasize openness, honesty, and care for each other. Additionally, these moments of inequitable status did not undermine senses of connection and support. Interactions between April, Kelsey, and Boden were substantially valuable, and in moments such as April's experience, Stephanie's role was supplemental to Kelsey's.

Conversely, each of us also recalled instances on this journey when others worked to abuse authority and left us feeling inadequate: the condescending legislator dismissing Stephanie's experiences; the scholar dismissing Boden's research and using graduate students for his research labor. A pedagogy of tenderness cannot erase such abuses of authority that "can shut people down, close their minds, and reinforce their preconceived ideas" about themselves and others (Thompson, 2017, p. 5). However, it does enable us to see these moments as ones steeped in power dynamics and inequalities, and to "give students a real experience of talking as equals across divides they were taught to uphold" (p. 5). Speaking across these longstanding and seemingly unassailable divides helps us to humanize complex and intimidating academic experiences, empowering us to make informed decisions to move through power structures to build new, more equitable connections.

Within these human moments of confusion, excitement, fear, frustration, and support, we find ourselves arriving at a shared/individual identity of "I am a qualitative methodologist." The two words, "qualitative research," intimidated all four of us at our start(s), but this community that we built within/because of qual have shifted those words from fear-inducing to emboldening and invaluably connecting. It is perhaps fitting that the disruptive and humanizing qualities of qualitative research are what initially excited all of us, and that our relationships have taken on similar qualities. It is through these and other relational moments that we individually/collectively find that our stories do count, and through one another we find "bigger ways of understanding ourselves" within and beyond research and education (Thompson, 2017, p. 6).

Mentorship in the now

April, fall 2020: A little over a year later, and I am officially a PhD student. I am in the process of transitioning from full-time employment to full-time student, and once again I find myself experiencing a mixture of anxiety and excitement. I have been working full-time and living independently for 15 years, so the prospect of not having that financial security gives me pause. However, the opportunity to submerge myself in qual, to experience it fully, floods me with gratitude and eagerness. I find myself desiring more opportunities to learn from others and to engage in mutually productive/beneficial relationships. I have heard how isolating PhD studies can be, so I want to be intentional about staying connected to those who are experienced travelers on this journey. This desire to be more relational is new for me, but I know that this intra-connectedness is a necessity if I am going to matriculate through the PhD program in a positive and healthy way.

Boden, summer 2019: I've just switched my doctoral program to qualitative research. My excitement about the future is tempered with sadness; I'm reeling. Dad passed away in September 2018 after a prolonged illness. My best friend Nick passed unexpectedly in June 2018. I haven't healed.

It's May 1. Stephanie emails me. She's editing a book entitled *Narratives of Hope and Grief in Higher Education*. I'm going to be copy-editing the book as part of my assistantship, but she's invited me to write a chapter, if I wish. I glance over the details: 3,000–5,000 words . . . grief . . . hope . . . higher education . . . draft by August 1.

During my dad's illness I had had two courses with Stephanie, and we had spoken throughout the experience. She had also lost her father to extended illness while completing her doctoral program. She understands my struggle. She knows that there's no leave-of-absence bereavement available to graduate students: Courses must be taught, grades must be assigned, assistantships must be carried out. I *need* to research; I can't have a line on my vita reading, "Dad and best friend died, couldn't work."

My ability to continue in my program is directly related to the extraordinary humanity of the program's faculty. They allow me to continue at my own pace. The last two semesters have been a blur of shifting due dates and make-up papers. Now it is summer, which is supposed to be my break; a time to pick up the pieces. I re-read the email. Summer is not for writing a book chapter. I initially agree to write the chapter, but over the next month I draft at least five emails to Stephanie explaining why I can't complete the project: I'm too tired, I don't want to relive this trauma. The book centers on grief *and* hope; I surely feel the grief, but I can't find the hope.

Still, I trudge ahead. I struggle. I converse with myself as I type. I talk on the phone with my brother, and we retell old stories. It is therapy. I also find meaning in Deleuze and Guattari's *A Thousand Plateaus*. Deleuze's discussion of "the refrain" helps to slow the chaos around me, to write my memory, to articulate why Dad meant so much to me. I write a story worth telling, about the most important man in my life. I'm so glad that Stephanie understood my grief, and I'm so glad that I didn't send any of those drafts.

Kelsey, spring 2020: Stephanie and I are co-teaching our Qual I class, struggling a bit in a discussion differentiating between positionality and subjectivity. Stephanie is now offering a third attempt to explain the difference, and I'm desperate to help because I remember struggling through this very concept in Qual I myself. Stephanie finishes her third iteration, and there's silence full of palpable apprehension. No one wants to ask us for yet another explanation, but it's clear that it just hasn't clicked yet. *Elevator speeches*, I think, remembering an exercise I've done a thousand times with the very professor with whom I'm co-teaching, and I take a stab at it. "Your positionality is who you are," I say, "your subjectivity is how it affects your research." A short pause. "There," exhales Stephanie, "that's a better way of putting it, say that again." I do, and the sound of typing/scribbling fills the room.

The "beginning" of it all: Qual II

Stephanie, fall and spring 2018: It was only because of outstanding mentorship that I even found qualitative research, and it was certainly because of several exceptional mentors that I learned to conceptualize myself as a methodologist. Each time that I meet with students, I remind myself of how utterly clueless I felt, and how frustrating it was to occupy a space of not-knowing when I was supposed to be smart, supposed to be earning a doctoral degree. I met each of my three brilliant co-authors when they enrolled in one of my Qualitative Research II courses, and so Qual II became an important cornerstone of my relationships with each of them. While Boden and I moved through spring 2018 together, April and Kelsey were both enrolled in fall 2018; that course proved invaluable to them and me, personally and academically.

Kelsey and April, entangled, fall 2018: We're about a third of the way through our Qual II semester, and one evening Stephanie asks us to find discussion partners. Specifically, we should pair up with someone with whom we haven't talked much. I make my way to the opposite side of the room; April and I haven't been partners yet, but that's mainly due to the fact that we sit the furthest away from each other.

I'm intimidated by April. She's the type of person who doesn't say much, but when she does, it packs a punch. She's honest without being combative, and I've noticed that when she offers an opinion, it usually aligns pretty closely with my own. I'm intimidated by her, but I want to be friends with her; it's an odd sensation.

I'm a little nervous sitting down next to her. She always seems very put-together, and I'm worried she'll see through me, right to the heart of my imposter syndrome. As we begin our discussion, though, my nerves calm. We share similar perspectives, and our conversation about the readings comes easily. She *is* very put-together – about that I wasn't wrong, but she's also very open about her own struggles as a student. Instant friendship overtakes the intimidation, and April and I are able to have some of the most challenging, *raw*, and introspective conversations I've ever had in a doctoral class.

April and Kelsey, same time/place: Kelsey and I have never shied away from difficult conversations. From feminism, politics, and educating family members on politically correct language – our chats often result in laughter, eye-rolling, and exasperation at others' obtuseness. It is *this* conversation on critical race theory (CRT) and Kelsey's research goals that shift our relationship; I move from mentee to mentor. Kelsey is wrestling with *who* gets to engage in CRT-based research, using words like "not qualified" or "worthy" to explore her own whiteness. As a Black woman who's invested in honest examinations of race/racism, I ask, "Why can't you engage in CRT-based research? If your intent is genuine, use the framework that aligns best." Although my words are not awe-inspiring, Kelsey's expression suggests that my feedback has offered a clearer path forward. Yet again, relationships have offered support and empowerment, and helped us to navigate what had seemed insurmountable academic hurdles just moments ago.

Shifting to the now/future: relationships and mentoring

As we four thought through this chapter and shared various moments that mattered to our trajectories, a common refrain was, "I didn't think of this moment as 'mentorship' until now." Given the near-silence regarding explicit considerations of how mentoring matters methodologically, our reflections helped us to realize that one reason for that may be the near-invisibility of those moments. Each of us has developed as a scholar through our connectedness. For example, and to emphasize the importance of mentorship that recognizes but works to disrupt hierarchy, Stephanie is a better researcher and instructor – because of Kelsey's clear explanations in class, April's many questions, and Boden's willingness to share his experiences. Faculty mentorship is unquestionably valuable, but when mentorship gets discussed, that is the element regularly highlighted. Our stories accentuate the degrees to which peer-to-peer mentorship matters, and the ways that faculty might defer to that form of mentorship. The point of mentoring is to foster independence and encourage growth, and there are moments when students present as more effective mentors – to one another and to faculty – than more traditional notions.

Mentorship as we have experienced it is complex; it is simultaneously/ contradictorily relational *and* independent. Unquestionably, none of us would have arrived at our current identity of "methodologist" without the other three, but those entanglements also support individual growth. Each of our research trajectories (dis)entangle into new paths that involve new people, new thoughts, while always tugging along threads that leave us tied to one another. Additionally, those threads spread out and, in new tangles, weave in new people. We close with Kelsey's reflections on how, as she began to assert her methodologist self beyond academia, she necessarily looped in others and extended notions of mentorship even further than we had previously imagined.

Kelsey and Stephanie, spring 2020: "Great," I think, as I sneak a glance at my mother's face. "She's got that *this-is-going-to-be-a-trainwreck-but-I'm-humoring-her* look."

We're walking to Starbucks to meet with Stephanie, my advisor. Visiting for the weekend, my mother has heard so much about Stephanie since I changed doctoral programs last year that I thought they should meet. Judging by that look, though, this might be a disaster.

We walk into the coffee shop, and Stephanie is at a table, working away at her laptop, as I thought she might be. I introduce them, and I can tell that Stephanie's uber-approachability is cracking my mom's preconceptions of what a doctoral advisor would be like. We get our coffees: a cloud macchiato for my mom, an iced vanilla latte for me, and an unsweet black iced tea for Stephanie.

When I first suggested this meeting, I suspected that the interaction would go well, but I still worried that the conversation would be forced or become awkward; I mean, how often does a mother have a casual chat with her daughter's doctoral advisor? But, they get on like biddies at a beauty parlor. I sip at my latte, watching as my mom quickly recognizes and understands one of the biggest reasons that I had changed doctoral programs: Stephanie's mentorship.

"She is *awesome*," my mom says walking back to the car, "I see what you mean now. She's so open and honest; I really want to be friends with her."

"I'll send you her Twitter handle," I say, smiling.

This moment did not literally involve April or Boden, but it immediately brought to mind the ways that *all* of our lives/worlds have been irrevocably informed/changed. A pedagogy of tenderness understands the importance of "those spontaneous, planned, and found" moments that remind us "that we all belong together. A willingness to travel together, to reach into the mysterious, the unknown," unafraid because of the relationships that guide and support us (Thompson, 2017, p. 5). In our travels, our academic relationships have intertwined with personal lives: Boden's brother helped him to work through Stephanie's book chapter invitation; April's mother demanded a copy of an article that she co-authored with Kelsey and Stephanie; Stephanie's family hears about April, Kelsey, and Boden regularly. This connected journey is one that emphasizes the degrees to which mentorship is inextricably bound into/with relationships. These communities that we build are "a sacred space, that [helps us to] coach each other" in ways that are powerful, agentive, and meaningful (Thompson, 2017, p. 5). For all of us, together/individually.

References

Brown, A. (2008). *Real talk: Lessons in uncommon sense: Nurturing potential & inspiring excellence in young people*. iUniverse.

Chawla, D., & Rawlins, W. K. (2004). Enabling reflexivity in a mentoring relationship. *Qualitative Inquiry, 10*(6), 963–978. https://dx.doi.org/10.1177/1077800404269420.

Cobanoglu, A. A., Yucel, Z. E., Uzunboylar, O., & Ceylan, B. (2017). A blended mentoring practice for designing material as a foreign language learning. *Turkish Online Journal of Qualitative Inquiry, 8*(1), 141–160.

Coppin, R., & Fisher, G. (2016). Professional association group mentoring for allied health professionals. *Qualitative Research in Organizations & Management, 11*(1), 2–21. https://dx.doi.org/10.1108/QROM-02-2015-1275.

Dönütü, O. U. Y. A., & Uygulamalar, N. N. (2020). Reflective peer feedback in the practicum: Qualitative and quantitative practices. *Turkish Online Journal of Qualitative Inquiry, 11*(1), 85–109.

Gunuc, S. (2015). Implementation and evaluation of technology and a mentoring program developed for teacher educators: A 6M-framework. *Qualitative Research in Education, 4*(2), 164–191. https://dx.doi.org/10.17583/qre.2015.1305.

Hoskins, M. L., & White, J. (2013). Relational inquiries and the research interview: Mentoring future researchers. *Qualitative Inquiry, 19*(3), 179–188. https://dx.doi.org/10.1177/1077800412466224.

Sassi, K., & Thomas, E. E. (2012). "If you weren't researching me and a friend": The mobius of friendship and mentorship as methodological approaches to qualitative research. *Qualitative Inquiry, 18*(10), 830–842. https://dx.doi.org/10.1177/1077800412456958.

Schulz, L. L., Tiner, K., Sewell, D., & Hirata, L. (2011). Towards a pedagogy of tenderness: Reflections on Cuban education. *International Journal of Humanities and Social Science, 1*(10), 55–58.

Thompson, B. (2017). *Teaching with tenderness: Toward an embodied practice*. University of Illinois Press.

MENTORING AMALGAM

Rachel K. Killam

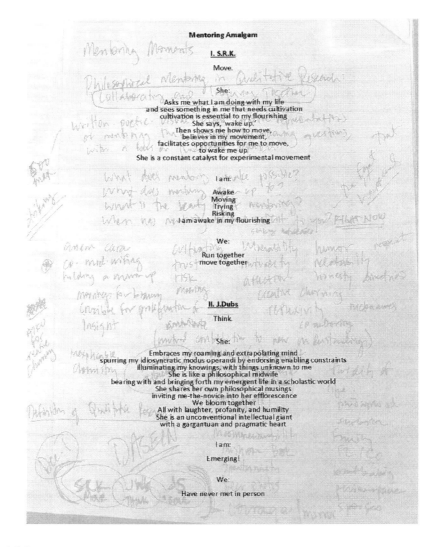

Mentoring Amalgam

I. S.R.K.

Move.

She:
Asks me what I am doing with my life
and sees something in me that needs cultivation
cultivation is essential to my flourishing
She says, 'wake up.'
Then shows me how to move,
believes in my movement,
facilitates opportunities for me to move,
to wake me up.
She is a constant catalyst for experimental movement

I am:

Awake
Moving
Trying
Risking
I am awake in my flourishing

We:
Run together
move together

II. J.Dubs

Think.

She:
Embraces my roaming and extrapolating mind
spurring my idiosyncratic modus operandi by endorsing enabling constraints
illuminating my knowings, with things unknown to me
She is like a philosophical midwife
bearing with and bringing forth my emergent life in a scholastic world
She shares her own philosophical musings
inviting me-the-novice into her efflorescence
We bloom together
All with laughter, profanity, and humility
She is an unconventional intellectual giant
with a gargantuan and pragmatic heart

I am:

Emerging!

We:

Have never met in person

DOI: 10.4324/9781003022558-19

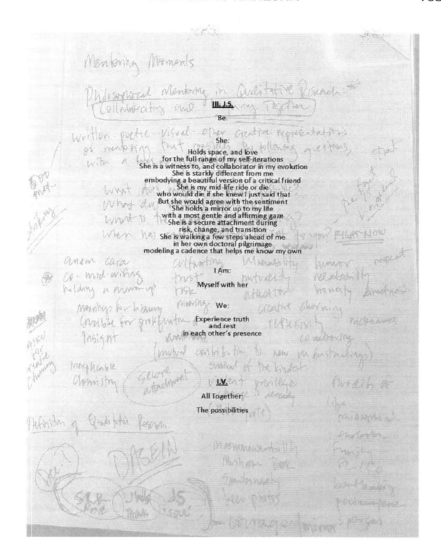

Mentoring Moments

Philosophical Mentoring in Qualitative Research:
Collaborating and _____ Together

III. J.S.

Be.

She:
Holds space, and love
for the full range of my self-iterations
She is a witness to, and collaborator in my evolution
She is starkly different from me
embodying a beautiful version of a critical friend
She is my mid-life ride or die
who would die if she knew I just said that
But she would agree with the sentiment
She holds a mirror up to my life
with a most gentle and affirming gaze
She is a secure attachment during
risk, change, and transition
She is walking a few steps ahead of me
in her own doctoral pilgrimage
modeling a cadence that helps me know my own

I Am:
Myself with her

We:
Experience truth
and rest
in each other's presence

I.V.

All Together:
The possibilities

MENTOR AND ATHENA

Minkyu Kim

< < < < > > > >

And Athena handed down her pacts of peace between both sides for all the years to come – the daughter of Zeus whose shield is storm and thunder, yes, but the goddess still kept Mentor's build and voice.

– Homer (1996, p. 485)

< < < < > > > >

In a recent interview with a research participant, a high school student, I asked what she found so appealing about a certain Korean pop music group that she likes so much. Her response ranged from their music to the concepts behind their music to the way they challenge gender stereotypes, which was important to her as she grapples with her own identity. When I asked her to say more about how they challenge gender stereotypes, she pointed out that they don't necessarily do it to make a statement, but rather, "they do it because that's how they are."

In thinking about the role of mentorship, we might recall that in *The Odyssey*, Mentor is a character's name; it is not his title. In other words, it is not an individual living up to a role. Odysseus entrusts his son (Telemachus) to Mentor because of who Mentor *is*: wise and virtuous. My initial impulse for this piece was to write about the professors in my program, who are involved in their own research and are people whose work is inspired

 DOI: 10.4324/9781003022558-20

and inspiring in its curiosity and virtue. It's hard to say how much I learn those things from their direct instruction. I think it's more accurate to say that that's just how they *are*, and I aspire.

However, I might also point out that most of Mentor's advice in *The Odyssey* does not actually come from Mentor himself. Rather, it comes from the goddess Athena, who borrows Mentor's "build and voice" to encourage Telemachus to embark on a research project of his own – namely, to find out what happened to his father.

I think there is a sense in which we think we know who we should turn to for knowledge: students to professors, for instance. Often those sources are indeed edifying and fruitful. But I think what we as education researchers are collectively arguing on some level is that we should also be open to lessons from places we sometimes overlook: in particular our young people, from whom life, wit, joy, and justice burst forth, unbridled and unmediated by adulthood and its attendant years of bureaucratic compromise and capitulation. In this analysis, the academic research institution is the build and voice of Mentor, what with all our source-citing and data-filled charts. But emanating from us is a voice beyond ourselves – a young music fan struggling with her identity, for instance, teaching us where one might instruct others to look for comfort – a sublime voice, bursting forth with storm and thunder.

References

Homer. (1996). *The odyssey* (R. Fagles, Trans.). Penguin.

8

"I DON'T WANT FEELINGS. I WANT TACOS."

TOWARD A NEW MATERIALIST MENTORING PRACTICE

Brenda Sifuentez and Ryan Evely Gildersleeve

Introduction

The relationship between the authors, Brenda and Ryan, emerged in the academy; as advisor and advisee, our roles were pre-established from the beginning, given the archetype of graduate training. However, our mentoring practice, which we have called *becoming-mentoring*, is not restricted to the human-human interpersonal relationship just described. We (Brenda and Ryan) instead focus on moving beyond a mentorship rooted in the exchange of knowledge and skills to one that demonstrates that mentoring should not be restricted to human interpersonal relationships that center feelings and abilities. Instead, our *becoming-mentoring* practice emerges from the intra-actions with the materiality of the world around us. This chapter uses new materialism and assemblage theory to conceptualize mentoring as nuance that should include the broader context of human and nonhuman entities by bringing attention to the performative, scripted, and spatial aspects of mentoring. We do this by introducing new materialism and assemblage theory as philosophies that inform and guide our *becoming-mentoring* practice, by conceptualizing a multifaceted engagement that moves beyond the interpersonal and transfer of knowledge relationship. Our *becoming-mentoring* practice opens up mentoring beyond traditional mentoring accounts that view the mentor and the protege in either an academic or psychosocial domain. Instead, the practice

DOI: 10.4324/9781003022558-21

of *becoming-mentoring* can move beyond predetermined places, feelings, and intrapersonal characteristics, to open up the possibility to the agency of the material and discursive. In this chapter, we situate our *becoming-mentoring* in the context of qualitative research. It was in the doing of qualitative inquiry that we were able to engage in discussions of *becoming researchers* and *mentoring*. The materialization of qualitative inquiry brought together opportunities for new ways of mentoring to occur outside of academic buildings. We begin by discussing how assemblage theory and new materialist philosophies allow our *becoming-mentoring* to take shape through the materiality produced in the field, classroom, and discussion over tacos.

New materialism

New materialism seeks to rethink the relationship between the discursive and material productions of knowledge by critically engaging with matter's resilience (Connolly, 2013). In new materialism, matter is not theorized in a state of equilibrium; rather, matter exposes immanent relations that self-organize (Coole & Frost, 2010). Yet instability and volatility of matter are not the same and are always redefining and reassembling elements for new and unexpected forms (Coole & Frost, 2010). For example, a mentoring relationship is not stable and static. Rather than viewing mentoring as a relationship based on hierarchy and intrapersonal characteristics, mentoring can be conceived as an open complex relationship that includes materiality of the assemblage.

New materialism disregards dichotomies of human/nonhuman and realigns the "human" as a situated product that is part of the broader process of materialization (Gildersleeve & Sifuentez, 2017). The materialization views bodies as actors within a society that is materially real and socially constructed. The ontology of new materialism challenges notions of singularity; instead, it recognizes the plural dimensions and complexity of being. Using a new materialist approach to mentoring acknowledges that no one person or thing is in charge or directs what occurs in a mentoring relationship. The exploration of mentoring using new materialism allows for the examination of complex and changeable casualties of mentoring that can occur on a micro social system and structural levels.

Assemblage theory

Working from an ontology that moves beyond dualisms of subject/object, assemblage theory generally encompasses three features in relationship with each other. DeLanda (2006) discusses the first relational feature of an assemblage, that the systems and elements function as both the content and expression of the assemblage. Together, these systems and elements produce something unique (e.g., a situated identity). The second relational feature acknowledges that no situation is ever static. Therefore, it is always moving towards something else as components emerge into new becomings. This feature is known as deterritorialization/reterritorialization (DeLanda, 2006; Deleuze& Guattari, 1986/1998). The deterritorialization/reterritorialization signifies a pre-established investment of objects (Deleuze & Guattari, 1986/1998), such as the ascribed roles between professor and student. In a state of deterritorialization/reterritorialization, *becoming* occurs when there is a change within the assemblage. A student may take the lead as facilitator of the class, thereby displacing the professor as lead facilitator. Therefore, the dominant discourse is malleable yet is still recognizable in stabilizing ways, even as new assemblages are de-stabilizing power to protect a given territory. Power in assemblage theory is not seen as an oppressive force; rather, it has multiple and productive forms. Territory should not be conceptualized only as a physical place but can include the social, discursive, and material. The final feature of the assemblage is its materiality, as the materiality provides impressions and insights to a world that is constantly *becoming*. Bodies and other materials are relational; they have an ontological status that is produced through their relationship with other bodies, things, and ideas (Fox & Alldred, 2015). Assemblages occur as various actions and events create chaotic networks that can be further reassembled in different ways to produce new assemblages (Potts, 2004).

In assemblage theory, the concept of "subject" does not exist. Rather, it results from *becoming* that conveys the changes and capacity of an entity. These changes can be physical, emotional, or social (Fox & Alldred, 2015). *Becoming* is more than just altering one's capacity; instead, it is a production of multiplicities; it represents social production that is nonlinear and capable of bringing about a new unity. Assemblages can function as territories that are produced by the effects among relations (DeLanda, 2006).

The ontological status of all assemblages is relational; therefore, hierarchy ceases to be a priority (DeLanda, 2006). Ontologically flat assemblages of humans, materials, and expressions in day-to-day intra-action constitute assemblage territories that are always in (de)formation (DeLanda, 2006). The dimensions and utilization of co-functioning systems in assemblage theory recognize that no parts of the assemblage fit together in a uniform way by nature or origin. As elements assemble, they establish mentoring relationships. Viewing mentoring as an assemblage recognizes territories that seek to establish a binary relationship between mentor and mentee for the development of careers, guidance, and a transfer of knowledge. With this recognition, assemblage theory opens the material world to influence and direct the ways that mentoring can occur. For instance, the categories within mentoring are less stabilized and the material conditions in which mentoring takes place are not isolated from one another, nor is materiality excluded from the practices of knowing.

Mentoring as an assemblage

Conventional components of mentorship are typically associated with the human experience that produces specific ways of knowing and being. Traditional mentoring models are framed by the relationships among humans (Lunsford et al., 2017; Yob & Crawford, 2012). In conventional mentoring models, success is often attributed to a faculty member's ability to graduate students and facilitate job placements (Lunsford et al., 2017; Okawa, 2002). Humanist mentoring occurs with the expectation that a student becomes a successful professional. These mentoring models have a humanistic perspective that centers on how humans possess the ability to act with their choices and exert a unidirectional relationship. Departing from traditional mentorship notions, we (Brenda and Ryan) apply a new materialist and assemblage theory framework to explain our *becoming-mentoring* practice. We argue that our *becoming-mentoring* practice seeks to move beyond a relationship between people and feelings. Including materiality allows for the possibility of questioning and temporarily suspending the steadfast borders (e.g., academic and non-academic) to move past the artificial boundaries that seek to confine a predetermined relationship. In describing our *becoming-mentoring* practice, we move beyond traditional understandings of mentoring to open up mentoring that is blurred so that we can learn through and alongside one another.

Assemblage theory from the lens of DeLanda (2006) and informed by Deleuze and Guattari (1980/1988) states that an assemblage entails examining the expression and context of the assemblage and the ways that deterritorialization and reterritorialization can create new *becomings*. Components of an assemblage have distinctive discreet assemblages that multiply as new lines of flight are created (DeLanda, 2006). Lines of flight are shifts in trajectory; this occurs when there is a shift in trajectory of a narrative that escapes the line of power (Deleuze, 1995). The interaction of various lines of flight creates the ability to act; this occurs because it comes from within the relationship rather than outside the relationship. In our *becoming-mentoring* practice, we are not bounded individuals but are situated products that make up a collection of assemblages. Assemblages are an individual entity such as a person, institutions of higher education, or a city, each having their own historical identity (DeLanda, 2006). As situated products, the collection of assemblages we intra-act with bring forth their own historical identities. We are more than human; we are a collection of the materiality that comprises us. Emergent lines of flight produce these components of assemblages. The breaking away from binary concepts and roles in new materialism and assemblage theory reveals that parallel outcomes can occur and, at times, create contradictory conclusions. What we gain when this happens is a new understanding of ethics and justice, not as something that is predestined, but rather as something that is continuously evolving. *Becoming-mentoring* is the refusal to directly center humans and their achievements. *Becoming-mentoring* instead captures the production of new multiplicities that among human and nonhuman entities produces a new understanding of mentoring that is always in flux.

In the following sections we seek to demonstrate how the world's materiality intra-acts to shape the various lines of flight to produce *becoming-mentoring* practices. Intra-action is the recognition that all things are relational not just when acted upon by agents (Barad, 2007). The assemblage of mentoring provides us the opportunity to excavate how mentoring is just one assemblage within a larger assemblage of higher education. Every intra-action with the surrounding environment, whether in the field or in the classroom, continuously and as situated products, informs the *becoming-mentoring* process. Interactions with the materiality of various settings such as the classroom, office, and research site create chaotic connections in which the emergence of new lines of flight are produced. In the following section, we will explore the materiality that was produced to describe the mentoring assemblage that has influenced our *becoming-mentoring* relationship.

Becoming-mentoring: materiality produced

Our *becoming-mentoring* practice emerges through the deterritorialization of the conventional mentoring assemblage, which allows for the expression and acknowledgment of nonhuman entities and objects. As lines of flight emerge and intrasect with one another in the mentoring assemblage, the practice of *becoming-mentoring* manifests itself temporally and spatially. In the following sections, Brenda and Ryan share three different mentoring events that we hope illuminate our *becoming-mentoring* practice. Two of these mentoring events are from our individual perspectives, and one is our combined perspective. Although we are sharing three events that emerged in our *becoming-mentoring* practice, this practice is never static. It is continuously evolving and producing new assemblages that are continually being de/reterritorialized.

Tacos, not feelings

Brenda's turn: As I walked into the restaurant with Ryan, the sounds of regional Mexican music blasted from the speakers, drowning out the sounds of other patrons talking among themselves, the banging of pots and pans in the kitchen, and the cooks discussing orders "los tacos de lengua llevan cebolla." The smell of Mexican food serves as a material reminder that we have left the campus field site. I looked at Ryan and smiled, knowing that we would partake in some delicious tacos. As Ryan and I walked up to the counter and reviewed the menu, we began to order after a few minutes. We had selected tacos de asada, al pastor, chicharron, chorizo, and agua de jamaica for lunch. After settling in, we began to debrief our morning field visit and the two interviews conducted in the morning. We arrived at this restaurant because our research on ritual cultures at Hispanic-Serving Institutions (HSIs) brought us to two HSIs in California. We agreed that we had learned new things about the institution, people we needed to follow up with, and ways to begin to see what the data was telling us. New materialist ontology informed our inquiry; we began to discuss how data was making itself logical to us (MacLure, 2013). The restaurant's materiality influenced the conversation's direction as it moved away from research to sharing more personally. At one point, I started taking out the onions of my tacos, and Ryan asked, "Why are you doing that? The onions make the taco," and I proceeded to share my story about how my family worked in the onion fields, which caused me to hate the smell and taste of onions. In this scenario, the taco's ingredients served as the materiality that shaped our *becoming-mentoring* assemblage's content and expression.

No longer confined to Ryan's office or even the city where our institution of higher education resides, we could shed the formal roles of mentor/mentee in a place that typically is not considered an academic setting. The restaurant's layout removed the academy's monotonous routines to allow our *becoming-mentoring* relationship to be malleable. The "there-ness" that is typically found in academic spaces where mentoring occurs shifted so that the content of our *mentoring-assemblage* was not the tedious transfer of skills. Instead, the *becoming-mentoring* produced new "there-ness." This new "there-ness" was the ability to have conversations about onions and family history that might not have occurred in formal academic spaces. At this moment, the materiality of the restaurant produced new ways to engage. Agency enacted by the taco directed our conversation to include family history and personal preferences creating something new.

We established an ethic of care (Noddings, 1984) for our *becoming-mentoring* that was not rooted in academic knowledge production. Nor was it rooted in the emotions of my family history. As I shared my family history with Ryan, I did not center the emotional experiences of working in the fields but instead served to establish our ethic of care for each other. As Noddings (1984) illustrates, caring is a practice out of obligation, not necessarily a natural act. As Ryan and I built our *becoming-mentoring* together, an ethic of care emerged necessarily because caring for one another mattered to the broader practices of knowledge production, specifically post-qualitative research practice. This ethic of care acknowledged my family and me as a situated product in the mentoring assemblage. We demonstrated an ethic of care for each other while simultaneously recognizing that we did not have to share our deepest emotions or feelings to respect each other. This is not to say that we could not share our deepest emotions or feelings, but rather that they were not necessary. Our care for each other existed regardless of what we chose to share with one another.

This mentoring moment challenged the expectations of what mentoring relationships can look like in the academy. I did not need to share feelings with Ryan to know that each interaction we have with each other is grounded in an ethic of care. The generic materiality of the academy had limited our ability to establish an ethic of care. For example, having a conversation regarding onions would most likely not occur in Ryan's office, but being outside the physical space and place of the academy allowed the materiality of the tacos to act upon us. This is not to say that this could not happen in the parameters of the academy. Still, it is one way that our *becoming-mentoring* practice is unscripted and temporal.

At this specific time and location, the ethic of care that we established moved beyond developing a traditional mentoring friendship; instead, it focused on the entanglement of our *becoming-mentoring* – an ethic of care that was not rooted in Brenda and Ryan's emotions, but our obligation to one another and to our *becoming-inquiry* practice. Our ethic of care is based not on our emotions or feelings but rather on a mutual understanding of situated products. Our ethic of care is an integral part of our *becoming-mentoring* practice. It acknowledges us as situated products, yet our approach does not center emotions and feelings. Rather, it centers the material practices that might be affected by emotions and feelings, but is not reliant on mutual engagement in such a private (i.e., personal) experience. In pursuing our new materialist practice, we amplify the discreetly material to produce a mentoring relationship that served us beyond the accounts of mentoring in conventional mentoring literature which often privilege a feeling/emotional dimension. We sought a mentoring relationship rooted in an ethic of care and privileging mentoring as a materialist practice.

None of this is to say that neither Ryan nor I experienced or expressed emotion or shared feelings with one another. We certainly did. However, neither of us expected the other to expect feelings or emotions to be shared. Our obligation was not to share, but rather to care.

Adding ingredients: research

Brenda and Ryan's turn: The transfer of research skills that often occurs while in the field takes place from mentor to mentee. This transfer of knowledge and skills is typically associated with conventional mentoring. Conceptualizing mentoring in the field as an assemblage shifts the ontological status of qualitative skills and knowledge held by the mentor. Assemblage theory flattens the ontological status in the mentoring assemblage, meaning the mentor no longer has hierarchy over a mentee. In the field we acknowledge the power differentials between us, yet recognize that the materiality surrounding us allowed us to move about the world differently. Our *becoming-mentoring* practice is a product of multiple lines of flight. It does more than transfer knowledge and skills from one person to the other – it too produces lines of flight as it intra-acts with other assemblages. In the following section, we will demonstrate how these assemblages are the research site's materiality, the technology used for interviews and field notes, ourselves, and participants as situated products. During field visits to HSIs, the ascribed identities of professor-student were temporarily suspended. We were able to work past the highly bounded aspects of conventional mentoring and qualitative research that center humans to encompass more-than-human aspects of inquiry.

As we (Brenda and Ryan) sat down on a bench that surrounded a tree at an HSI campus, we shifted our focus from the humans (e.g., students, faculty, administrators) to be more intentional about what we were hearing and seeing. What were the sounds of the campus? After a few moments, we began to share what we heard. We began to discuss how sounds were associated with campus life. On this particular HSI campus, construction had been ongoing for the past two years. By focusing on campus sounds, we moved away from the human to consider the materiality of the campus construction. Conventional qualitative inquiry focuses on the human through interviews. It seeks to make meaning, yet here, our *becoming-mentoring* practice sought to deterritorialize aspects of qualitative inquiry and mentoring by including the sounds of the materiality of campus life at an HSI. The materiality of the campus allowed for the distinctions between human and nonhuman to be blurred. Knowledge creation was no longer about the transfer from the mentor to the mentee; rather, we were learning alongside human/nonhuman entities.

As we moved beyond traditional data collection methods to include campus materiality, the deterritorialization of data collection was occurring. Our *becoming-mentoring* practice provided the space to focus on how the nonhuman entities comprised campus life and its rituals. Traditional interviews seek to obtain an authentic human experience; the *becoming-mentoring* practice created an opportunity to move beyond traditional data collection methods and understanding. The data collection that occurred exceeded the interviews we had conducted to include the sights and machine-created campus sounds. The discussion focused on what the sounds of machines told us about rituals and campus life. Our inquiry was informed and framed by thinking with theory (Jackson & Mazzei, 2011). As we shared initial thoughts of what the campus materiality was doing to us at that moment, we inquired what it could mean for our current and future research on HSIs. The practice of *becoming-mentoring* took into account the temporal, spatial, nonhuman, and material aspects of the campus by moving away from the established relations of transferring skills and knowledge between human entities.

We were able to be part of and think with the materiality of qualitative data that sought to further entangle our practice as *becoming-researcher* and *becoming-mentoring*. The mentoring assemblage's deterritorialization created a shift from the professor's deposit of knowledge to the student, to where we could acknowledge power structures yet still work through them. Power was instead fluid rather than unidirectional; the knowledge being produced captured the material surroundings. Our field notes became more than just paper, pencil, and words; instead, they became the material that did something to us. The production of field notes allowed us to materialize the assemblage at the HSIs. Our field notes and individual interpretations of the theory were already intertwined, yet our *becoming-mentoring* practice allowed for thinking with theory. We moved beyond traditional qualitative inquiry to envision rituals, sounds, and structures as the campus's materiality and assemblage. Amplifying the material of inquiry allowed us to produce a mentoring relationship that focused on an ethic of shared knowledge creation that would support our mutual productivity as *becoming-scholars*.

Adding ingredients: classroom

Ryan's turn: Returning to campus, where the hierarchical power relations between a professor and a graduate student are reified not only in title, but also in spatial relations (e.g., my office had a door; Brenda's cubicle sat outside of it), access to various places on campus (e.g., my key card allowed me entry to the faculty reading room; Brenda's did not), and the arrangement of material structures to support our work (e.g., I had a fixed, institutionally recognized mailbox in the workroom; Brenda had a DIY file folder mailbox on her desk). Yet, in a parallel outcome, the *becoming-mentoring* practice could emerge wherein we could conjoin our methodological development via our co-constructed classroom space – amidst the reified power relations that positioned myself higher than Brenda. I bring this example to the fore at this point in order to acknowledge that power relations exist in tension with emergent assemblages, as parallel outcomes. That is, multiple lines of flight can take off, sustain, and intersect one another while differential *becoming*-practices are engaged on multiple planes. Put simply, even though in many ways returning to campus highlighted power dynamics that never left our relationship, our *becoming-mentoring* practice could still emerge in a flattened ontological plane as a simultaneity.

In the classroom, our assigned roles were Teaching Assistant and Instructor-of-Record. Discursively, we were clearly in a hierarchical structure. To engage our *becoming-mentoring* practice, which relied on flattening our ontological status, each of us needed to take risks, engender trust, and demonstrate commitment to our shared interests in *becoming-mentoring* practice in order to produce knowledge around new materialism and post-qualitative inquiry. In reality, the power dynamics of our discursive positionings remained constant. Simultaneously, our *becoming-mentoring* practice together flattened our ontological status. Through our scholarship of teaching and learning, our *becoming-mentoring* practice relied, in part, on our ability to harness the affordances of our material environment and the materiality of our practice itself. In a sense, our *becoming-mentoring* practice was both a power-management tool as well as a parallel outcome to the discursive structuring of our relational positionings.

The classroom, like the Mexican restaurant where we enjoyed our tacos, greeted us with a cacophony of sounds, languages, smells of snacks and foods and bodies brought in by other seminar participants, and sights of technologies (i.e., screens, smartphones, laptops, pads of paper, pens, books, articles, etc.). While it is common practice for some instructors to check in with students about their emotional well-being and states of mind at the beginning of class meetings, we made prior choices to avoid this practice. This was not out of a lack of concern for one another nor for students in the class. We generally have found such practice to evacuate affect rather than amplify or harness the power of affect. We simply find greater growth in ourselves when attending to the doing of inquiry rather than the feeling of our extant lives. After all, we are engaged in a *becoming-mentoring* practice and leading a *becoming*-learning practice; we are not romantic partners, roommates, or otherwise beholden to one another. Rather, we chose to risk avoiding such evacuated constructs of feelings in preference for material relations and practices, which might include, but did not necessitate, the disclosure of affective relations. We instead inquired about what we each had done with the readings since the last class. What were we able to do? What helped us? What got in our way with what we sought to do with the readings? And we asked this on a whiteboard with markers – not a simple verbal request. And similarly to how we noticed the material doings of our tacos, we engaged with the materiality of doing the study practice of learning post-qualitative inquiry.

By preferencing the material relations of the study practice at hand, without necessitating that students, nor Brenda, nor I be required to engage in emotional labor together in the line of flight of our *becoming-mentoring* practice, we further contributed to a sense of trust between us. This trust had been established by knowing that we had an ethic of care and development in our *becoming-mentoring* practice. We each positioned ourselves physically in the classroom some distance apart, in order to take notice of similar things operating, yet from different vantage points. We relied on one another to see the taco-inspired pedagogy infused with life in the classroom. We necessitated support from one another in order to succeed in shifting the gaze of a class's beginning moments from feelings to doings. Such bricolage of instruction-cum-*becoming*-mentoring demonstrated our commitment to the *becoming-mentoring* practice that we sought to generate, yet without knowing where it might find us, or where we might be found.

Conclusion

Having new materialism and assemblage theory inform and shape the mentoring relationship that had been influenced by the ascribed roles of mentor and mentee allowed us to break away from the binary roles to flatten our ontological situated positions within these assemblages. Our relational mentorship sought to create something new through inquiry that went beyond interpersonal relationships of feelings, to produce something new.

The various lines of flight that emerge enable the potential for *becoming-mentoring* practices to be de/reterritorialized, widening participation outside of the human to include nonhuman entities and objects. The fundamental shift from only conceptualizing mentoring as between humans to that of *becoming-mentoring* opens up the possibilities of recognizing entanglements and assemblages that move beyond the transfer of knowledge and skills to examine new multiplicities of mentoring and research. Our *becoming-mentoring* practice moves beyond the linear nature of professional development. It is constantly evolving, shifting, and encompassing new assemblages. Our *becoming-mentoring* relationship was established in the field while conducting qualitative inquiry, specifically post-qualitative inquiry. The materiality that qualitative inquiry creates allowed for a new way of conceptualizing mentorship and research. The impact of our *becoming-mentoring* practice is yet to be known, and we may never know, as it will continue to respond to new emerging lines of flight that are immanently formed. Their formation generates new pathways that are both materially real and socially constructed, pathways that will transform the *becoming-mentor*.

References

Barad, K. (2007). *Meeting the universe halfway: Quantum physics and the entanglement of matter meaning*. Duke University Press.

Connolly, W. E. (2013). The "new materialism" and the fragility of things. *Journal of International Studies, 41*(3), 399–412. https://doi.org/10.1177/0305829813486849

Coole, D., & Frost, S. (2010). Introducing the new materialisms. In D. Coole & S. Frost (Eds.), *New materialisms: Ontology, agency, and politics* (pp. 1–46). Duke University Press.

DeLanda, M. (2006). *A new philosophy of society: Assemblage theory and social complexity*. Continuum.

Deleuze, G. (1995). *Negotiations* (M. Joughin, Trans.). Columbia University Press.

Deleuze, G., & Guattari, F. (1988). *A thousand plateaus: Capitalism and schizophrenia*. Bloomsbury Publishing. (Original work published 1980).

Deleuze, G., & Guattari, F. (1998). *Kafka: Towards a minor literature*. Ediciones era. (Original work published 1986).

Fox, N. J., & Alldred, P. (2015). New materialist social inquiry: Designs, methods and the research-assemblage. *International Journal of Social Research Methodology, 18*(4), 399–414. https://doi.org/10.1080/13645579.2014.921458

Gildersleeve, R. E., & Sifuentez, B. J. (2017). Latino/a youth activism in higher education: A new materialist analysis of the Latino graduation ceremony. *Critical Questions in Education, 8*(4), 342–357. https://academyedstudies.files.wordpress.com/2017/10/gildersleevesifuentezfinal.pdf

Jackson, A. Y., & Mazzei, L. (2011). *Thinking with theory in qualitative research: Viewing data across multiple perspectives*. Routledge.

Lunsford, L. G., Crisp, G., Dolan, E. L., & Wuetherick, B. (2017). Mentoring in higher education. In *The SAGE handbook of mentoring* (pp. 316–334). SAGE.

MacLure, M. (2013). The wonder of data. *Cultural Studies ↔ Critical Methodologies, 13*(4), 228–232. https://doi.org/10.1177/1532708613487863

Noddings, N. (1984). *Caring: A feminine approach to ethics and moral education*. University of California Press.

Okawa, G. Y. (2002). Diving for pearls: Mentoring as cultural and activist practice among academics of color. *College Composition and Communication, 53*(3), 507–532. https://doi.org/10.2307/1512136

Potts, A. (2004). Deleuze on Viagra (or, what can a Viagra-body do?). *Body & Society, 10*(1), 17–36. https://doi.org/10.1177/1357034X04041759

Yob, I., & Crawford, L. (2012). Conceptual framework for mentoring doctoral students. *Higher Learning Research Communications, 2*(2), 34-47.

CULTURALLY RELEVANT MENTORSHIP IN MOTION

Rouhollah Aghasaleh

Imagine a minivan packed with science teaching materials navigating what Hamann et al. (2002) call the "new Latino diaspora" in the South where long-term residents are "unaccustomed to outsiders of any type, or . . . have conceptualized difference only in dichotomous terms into which newcomers do not readily fit" (p. 2). The minivan commuted between 10 middle and high schools and other sites to provide professional development for almost 50 science and ESOL teachers and served 4,000 students and 100 Latinx families through multiple events, including a series of Steps-to-College through Science bilingual family workshops. Later, these efforts were recognized as one of the White House's "Bright Spots in Hispanic Education Fulfilling America's Future" (2015). Inside this vehicle, two leading scholars and several graduate research assistants – all international students – discuss their highlights from their interactions with Latinx families and identify strategies to support emergent bilingual youth, which quickly turns into discussions of high theories; including post-structural, new materialist, and feminist theories.

While this large-scale project[12] was designed to support immigrant students in secondary science learning, it had an unwritten goal to support international doctoral students from Uzbekistan, Turkey, Korea, México,

---

12 Language-Rich Inquiry Science with English Language Learners through Biotechnology (LISELL-B) www.nsf.gov/awardsearch/showAward?AWD_ID=1316398

DOI: 10.4324/9781003022558-22

Iran, Uruguay, China, Colombia, and Canada. This support was not limited to a financial one (although that was most definitely needed). More importantly, it provided a safe and prosperous environment to help them navigate the discursively, culturally, and politically complex path that emergent scholars had to travel. Our mentors did not make it explicit that they were aware of the cultural taxation and racism that we, as minoritized students, endured in and out of the institution. However, they went above and beyond what was expected from a mentor to support all of us in our personal and public lives. This thoughtful approach made us feel fully human, supported, and helped us to grow in our scholarly and personal capacities. This was at the time unlike most of our progressive teachers and friends who embodied benevolent racism as they tried to be our White Saviors.

There is a misconception among educators who embrace culturally relevant pedagogy when calls for inclusivity and equity are perceived as a call for leniency and lower standards. Our mentors never lowered the expectation because of the gap in our cultural and linguistic repertoires. They instead encouraged us to draw on our cultural capitals as assets to enhance our scholarship. An indication of their success is that all those young faces on this image are now accomplished assistant and associate professors at major universities in Massachusetts, South Carolina, Alabama, California, Georgia, Mississippi, and Michigan.

This mentorship never became salient until I graduated from my doctoral program and started working as an independent scholar. Although I was a strong-headed and opinionated student with a diverse (and sometimes scattered) range of interests, my mentors never told me what I *could not* do; instead, they showed me *how* to do it. I was given lots of autonomy to design, implement, and test what I had on my mind. Thus, I felt empowered, agentic, and important, a treasure for an *outsider* (which is what almost all other parts of my academic and everyday experience made me feel).

References

Bright Spots in Hispanic Education Fulfilling America's Future: White House Initiative on Educational Excellence for Hispanics (2015). Retrieved from https://www2.ed.gov/about/inits/list/hispanic-initiative/bright-spots.html

Hamann, E. T., Wortham, S., & Murillo, Jr, E. G. (2002). Education and policy in the new Latino diaspora. In S. E. F. Wortham, E. G. Murillo, & E. T. Hamann (Eds.), *Education in the new Latino diaspora: Policy and the politics of identity* (Vol. 2). London: Ablex Publishing.

A QUALITATIVE MENTORING PROCESS DIAGRAM IN HAIKU

Jaclyn Kuspiel Murray

A QUALITATIVE MENTORING PROCESS DIAGRAM EXPRESSED IN HAIKU

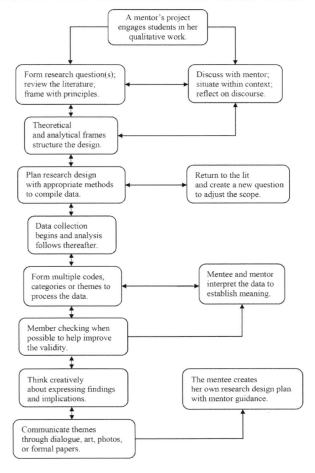

DOI: 10.4324/9781003022558-23

187

9

STRING FIGURE MENTORING IN A QUALITATIVE INQUIRY ASSEMBLAGE

Susan Naomi Nordstrom

Introduction

In this chapter, I articulate string figure mentoring, a way to conceptualize and practice qualitative inquiry mentoring in a territorial assemblage, by theorizing examples from my mentoring practice. To develop string figure mentoring, I considered the following two questions. What is the work of the term "qualitative inquiry mentor"? What do I do when I think I am mentoring someone in qualitative inquiry? When I think I am mentoring someone, I rarely do so as a sage expert who bestows knowledge upon a nescient apprentice. This would assume a hierarchical and binary relationship of expert/novice, one in which I impart my knowledge to a less experienced other (Buck et al., 2009; Doloriert et al., 2012; Engels-Schwarzpaul & Peters, 2013; Gordon, 2003; Knox et al., 2011). Instead, I spend a great deal of time listening and thinking with the person with whom I am conversing. I listen and think-with in order to learn from them and to foster generative connections that may very well lead to new thoughts. For me, the work of the word "mentor" and what I think I do when I mentor someone is much more a thinking-with assemblages, thinking-with connective relations, and thinking-with to generate new becomings.

In my mentoring practice, the relationship between expert and novice follows Deleuze and Guattari's (1980/1987) logic of the *and* in which the subject positions of expert and novice move and shift together to create new

 DOI: 10.4324/9781003022558-24

patterns, new ways of being, and new ways of doing qualitative inquiry. In many ways, these relational movements are what Haraway (2018) called string figures, a speculative and relational practice of "passing patterns back and forth, giving and receiving, patterning, holding the unasked-for-pattern in one's hands" (p. 61). As I think with people, doing the work of giving and receiving thoughts in order to create something new, I consider myself not a mentor, but a string figurer. A person engaged in Haraway's (2016) string figure work, thinking-with in order to create something new. To articulate this iteration of mentoring, I first think with qualitative inquiry by positioning the field within a flat ontology that disrupts binary divisions and hierarchies. This positioning then allows me to disrupt the expert/novice binary in qualitative inquiry mentoring and place it within a logic of the *and* (Deleuze & Guattari, 1980/1987). Following that, I develop string figure mentoring, a way to conceptualize mentoring within a flat ontology that aims to generate new becomings.

The failure of the big tent metaphor, the emergence of an assemblage, and qualitative inquiry mentoring

As a qualitative research methodologist in an educational research program that serves colleges across the university, I have seen the Big Tent of qualitative inquiry (Denzin, 2008) at work at my university. Denzin (2008) developed the metaphor in response to the early 2000s re-emergence of the paradigm wars following Scientifically Based Research in Education and the mixed-methods debates. This metaphor accompanied an offering of ten theses for how qualitative inquiry might operate in the new century. These theses work toward paradigm proliferation, openness to differing paradigms, recognition of paradigm choice as personal, dialog between paradigms, and a refusal of oversimplification of paradigm movements. Specific to the debates about mixed methods during the early 2000s, Denzin argued that "we must expand the size of our tent, indeed a bigger tent! We cannot afford to fight with another" (p. 321). While the metaphor was developed to offer a way to think through a particular historical moment, the metaphor of the big tent has gained traction as a way to describe the wide array of theoretical frameworks and approaches to qualitative inquiry (e.g., Roulston & Bhattacharya, 2018; Tracy, 2010). As a qualitative research community, we have generally accepted this metaphor and begun our attempts of living and working within it (Lakoff & Johnson, 2003). However, this metaphor no longer serves me, particularly in my mentoring work.

A big tent and a rubber balloon

Early on in my career as a qualitative methodologist, I learned that within one university (sometimes within a single department), multiple paradigms exist. For some faculty (and departments), the paradigm wars of the 1980s are still alive and well. For others, qualitative research (even in its most post-positivist forms) is still considered a fairly new and novel approach to research. Still for others, new approaches, such as post-qualitative approaches, are accepted as legitimate forms of knowledge production. As a qualitative methodologist, I have had to learn to honor each of these approaches, recognize that paradigm selection is both deeply personal and political, create dialogue between paradigms, and refuse oversimplification of paradigms (Denzin, 2008). I have had to learn how to do this work because disciplinary methodological politics shape what students perceive they can and cannot do methodologically in a study. I have learned these practices alongside students, walking alongside them as we navigate the big tent together. The withness of these movements have led me to question the metaphor of the big tent.

It is important to note that while qualitative inquiry is a discipline in and of itself, it is almost always done within and sometimes against other academic disciplines. Mentoring students across campus has allowed me to see how students and I must work together to navigate the political forces that emanate from both academic disciplines and qualitative inquiry. These forces, along with a student's personal desires, shape a student's project such that the project becomes a product of those forces. For one of his comprehensive exams, a student completed and published a Foucauldian (Foucault, 1984/1985, 1984/1986) interview study in which he drew on care of the self to design and implement a study about covert stuttering in the field of communication sciences (Constantino et al., 2017). While most programs offer an examination of coursework for comprehensive exams, this student's program area offered an article option in which students complete a smaller research study and then publish the study. His study was based on his readings of Foucault's (1984/1985, 1984/1986) care of the self and how much it personally helped him to think about his own stuttering practices. His design was very much informed by post-qualitative approaches such as Jackson and Mazzei's (2012) thinking with theory. However, his field is dominated by post-positivism. As the student, his major professor, and I worked on that study, we had to navigate why such an approach was valuable to his discipline and how the results of his study created more nuanced understandings of covert stuttering. While the metaphor of the big tent may be apt in this situation, the metaphor does not adequately acknowledge the political work we did.

Deleuze and Guattari's (1980/1987) ideas of reterritorializing and deterritorializing forces better explains the work we did. Working within and against – in the middle of – methodological, political, and personal forces created space to think about how these forces both reterritorialized and deterritorialized the study. In order to make the study intelligible to the field of communication sciences, we had to navigate these political forces with prudence. The student created a post-qualitative study while adhering to the key ideas of post-qualitative work (i.e., post-structural theories driving the inquiry process). However, the reviewers of the article were not well-versed in post-qualitative work, much less qualitative work. The student and I spent one meeting deciphering the reviewer comments about the article for these forces and then working on what to do with those forces. One reviewer appeared to have a limited understanding of qualitative inquiry in general and generated reterritorializing forces about the study that had to be navigated with political care. In responding to the reviewer comments, we had to adhere to the study of the design but write it in such a way that it would be considered "rigorous" and "valid" by a reviewer who appeared to have limited understanding of qualitative design, much less a Foucauldian-designed study. Navigating competing reterritorializing and deterritorializing forces became an integral part of finishing work on this article and getting it published.

A person's departmental methodological politics also figure into these forces. For example, the student had taken the courses required for our graduate certificate in qualitative research. In addition, he and I read Karen Barad's work together in an independent study. This would not have been possible without the support of his advisor and home department. This particular advisor and home department were open to qualitative research writ large. Other departments on our campus are less welcoming to qualitative inquiry. The mentoring I do across campus must navigate at least three political forces – qualitative inquiry politics, the discipline's methodological politics, and departmental methodological politics – in addition to the students' personal desires for their scholarship. Consequently, students and I find ourselves working and thinking with numerous political forces as we collaborate to produce knowledge.

If anything, the mentoring work with this student and others gestured not toward the big tent (Denzin, 2008), but rather a Bergsonian (1920/2002) reality that "resembles a gradually expanding rubber balloon assuming at each moment unexpected forms" (p. 226). In other words, this particular study, this unexpected form, stretched the rubber balloon in different ways. These unexpected forms make the balloon groan and stretch such that it may not even resemble a balloon anymore. Perhaps, it will even burst. In this example and others, the forces at work in the study overwhelmed the tent and the balloon. The metaphor began to lose its descriptive capacities. Consequently, I have begun to conceptualize qualitative inquiry as a territorial assemblage within a flat ontology.

Thinking-with a qualitative inquiry territorial assemblage

It is within a "Deleuzoguattarian" flat ontology that operates within the "logic of the and" (1980/1987, p. 25). Deleuze and Guattari wrote:

> Between things does not designate a localizable relation going from one thing to the other and back again, but a perpendicular direction, a transversal movement that sweeps one and the other away, a stream without beginning or end that undermines its banks and picks up speed in the middle.
>
> (p. 25)

In other words, the middle is in between multiple forces that are multi-directional, have neither beginning nor end, and are shifting. The "and" becomes useful to work within and against these forces. Returning to the prior example, the student, the other faculty member, and I were working in the middle of his home discipline's forces, post-qualitative forces, his chosen theoretical framework, reviewers, and . . . and . . . and. From these ands, the work picked up speed as it became transformed.

Qualitative inquiry as a flat ontology offers a way to think through all these reterritorializing and deterritorializing forces and the politics associated with those forces. These generative forces create iterations of qualitative inquiry. Conceptualizing qualitative inquiry in such an ontological framework is useful in that we can begin to see the field itself as something that is always in a state of creation. In such a generative ontological state, qualitative inquiry becomes a series of iterations that are contingent manifestations of personal, political, ethical, disciplinary, and other forces. As qualitative inquiry continues to materialize these unexpected forms that expand the field in unanticipated ways, qualitative inquiry functions as a territorial assemblage.

Deleuze and Guattari (1980/1987) wrote, "The territorial assemblage is inseparable from lines or coefficients of deterritorialization, passages, and relays toward other assemblages" (p. 333). All the lines that have deterritorialized conventional qualitative practices (e.g., feminist, decolonizing, indigenous, post-qualitative, and so on) are within the qualitative inquiry territorial assemblage. These lines carry with them all the politics that generated their materialization within the assemblage. The qualitative inquiry assemblage links to an innumerable amount of other assemblages such as other academic disciplinary assemblages and departmental assemblages, all with their own set of deterritorializing and reterritorializing lines. As these assemblages connect, disconnect, and reconnect, new iterations of qualitative inquiry are added to the qualitative inquiry assemblage.

By conceptualizing qualitative inquiry as a territorial assemblage, it becomes possible to shift Denzin's (2008) argument about the big tent to a flat ontology that generates a supple, contingent, and shifting qualitative inquiry assemblage that is far from singular. Denzin's (2008) theses become entangled vectors that must be worked through as they create dialogue that recognizes and honors the personal and political work of qualitative inquiry. This assemblage is responsive to other disciplinary assemblages, departmental assemblages, and other assemblages. The movements of these assemblages become sites of negotiating political lines that seek to both reterritorialize and deterritorialize qualitative inquiry. These negotiations bend and shape what is both possible and impossible within the qualitative inquiry assemblage. As qualitative inquiry continues to grow and materialize, qualitative inquiry creates itself anew, transforms itself, as new deterritorializing lines of flight are generated within the assemblage.

Experts and novices in a qualitative inquiry territorial assemblage

Conceived in this way, a qualitative inquiry assemblage creates different mentoring practices that in turn create different becomings of expert and novice. Mentoring assumes two subject positions: expert and novice. In effect, the subject positions, expert and novice, assume a pre-given stable subject position. Specific to qualitative research mentoring, it is sometimes assumed that the mentor (the expert) has more extensive knowledge in qualitative inquiry and shares that knowledge with the mentee (the novice). The subject positions are usually granted by years of experience or particular past experiences (e.g., moving through tenure and promotion and publishing articles) within an academic field. Simply put, an expert is someone who has gone through whatever process the novice has not. The expert has knowledge about that process and imparts it on the novice. Thus, a binary materializes – expert/novice. The literature on mentoring in higher education is dependent on the binary relationship of expert/novice (e.g., Gordon, 2003; Knox et al., 2011). Conceptualizing the field of qualitative inquiry as a territorial assemblage, however, disrupts these subject positions and the binary relationship that materializes between those subject positions.

The binary of expert/novice assumes a hierarchical power relationship in which generally, the expert has more power than does the novice. Literature on mentoring acknowledges this hierarchical relationship, suggesting that this hierarchical arrangement warrants a reflexive approach (Buck et al., 2009; Doloriert et al., 2012; Engels-Schwarzpaul & Peters, 2013). While these authors did not operate within a Deleuzoguattarian framework, thoughtful work about power must be done within assemblages. Assemblages are not hierarchical. Instead, power, much like Foucault's (1976/1978) conceptualization of power, is always moving with deterritorializing lines of flight and reterritorializing lines that seek to stratify hegemonic ideas. Power relations are always moving and shifting and should be allowed to move and shift – a critical move in mentoring in assemblages. Returning to the example in the previous section, the article in question was part of the student's doctoral comprehensive exams. As a member of his committee, I had to sign off on whether he passed his comprehensive exams, the other exams included. To be quite honest, I had to rely on the other committee members' expertise to know how he performed on the exams, one of which included neuroscience (a subject in which I have no expertise). Likewise, the other committee members relied upon my expertise to know how he performed in the article. The student was in the middle of our varying areas of expertise and many times had to translate areas of diverse knowledge practices to the committee members. While the committee members may have had expertise in their respective subject areas, the student was an expert in the intersection of these subject areas. This novel intersection of subjects was produced by this specific mentoring assemblage. Power was flowing as no one person on the committee was fully an expert or novice. These flows destabilized the binary of expert/novice and put them into a logic of the *and* (Deleuze & Guattari, 1980/1987).

Here it is useful to think with Stengers (2010/2011) and her work with nomadic and sedentary individuals. She argues that these subject positions can be identified only "in relation to a given interaction, of creating a contrast whose scope does not exceed that interaction" (p. 364). That interaction happens within "a territory that assigns limits and conditions to the risks they boast of" (p. 364). Likewise, the assemblage in which an "expert" and "novice" are situated creates conditions that make possible certain interactions and, in turn, create different iterations of expert and novice. Returning to the example, each interaction I had with the student created different iterations of expert and novice. In some instances, he became more expert in terms of navigating his own discipline, and I became more novice. In other instances, such as responding to reviewer comments on the article, he and I became both expert and novice as I shared ways to respond to reviewer comments, and he navigated the disciplinary politics. We were never fully expert or novice. Instead, the mentoring relationship within the assemblages in which we found ourselves generated degrees of both expert and novice.

While Watkins (2010) described the affective relationship between teacher and student, it is useful in thinking about mentoring. They wrote, "the relationship between teacher and student may not be an equal one but its success depends upon mutuality, a recognition of worth by both parties with this intersubjective acknowledgement being integral to their sense of self" (p. 73). In other words, the teacher is expert, the student is novice, and those roles must be appreciated in conventional mentoring. When "there exist no other drives than the assemblages themselves" (Deleuze & Guattari, 1980/1987, p. 259), there can be only becoming both expert and novice. Becoming is about creating and transforming heterogeneity within the course of events (Deleuze & Guattari, 1980/1987). What I have hoped to demonstrate with the extended example with the student throughout this chapter has been the following. As events passed through the student and me in the mentoring relationship, different power relations materialized that in turn created different iterations of expert and novice such that we were always already becoming both expert and novice. The degrees of becoming both expert and novice are responsive moments in the mentoring relationship. In turn, the relationship becomes a series of movements that both parties must acknowledge and respect to keep power flowing in generative ways.

These relationships, however, do not only create degrees of becoming expert and novice. In assemblages, one cannot simply divorce themselves from other identities. For example, during analysis, the student and I frequently drew on our personal experiential expertise to work through the various ways Foucault's (1984/1985, 1984/1986) care of the self concepts moved through the data. The student is a stutterer, and his experiential knowledge was critical during analysis. As a woman, I was able to draw on my experiential knowledge as a woman to assist with data relating to gender. Upon reflecting on analysis, he and I discussed how our personal experiential expertise figured heavily into our becoming expert and novice together as we listened to and learned from our experiential knowledge. Simply put, we were never just two experts and novices. We always already were and are expert and novice within the array of identities that constitute our becomings.

These relationships are generated by events that operate in a coexistence of time. Reflexivity, as it is practiced in the aforementioned literature about mentoring in higher education, is dependent upon a conceptualization of a time in which past, present, and future are delineated. However, in assemblages, pasts, presents, and futures are happening in a coexistence of time (Deleuze & Guattari, 1980/1987). Assemblages and the events that constitute them create a series of nows that are past-present-future moments that generate degrees of both expert and novice in mentoring that are always already materializing in subjects. In other words, no one can be fully expert or fully novice. One is "both and" at all times as events pass through them and past-present-futures animate possible degrees of both expert and mentor. The student and I have kept in contact since his graduation in 2017. In fact, he served as a critical reader of this chapter and provided valuable feedback. For example, his reading helped me to see what I am trying to achieve in this chapter, a new telos of mentoring that is generated from active becoming-with each other. Each event that carries with it its own past-present-futures continues to create new iterations of our mentoring relationship.

Degrees of becoming both expert and novice materialize from assemblages that pulse with the logic of the *and* that refuses binary relationships (Deleuze & Guattari, 1980/1987). Assemblages are about movement and keeping the assemblages in a state of movement. To do this work, one must not reterritorialize either the expert or novice into a pre-given subject position, which would indicate a practice of hierarchical power relations. Moreover, one can never master all the potential forces in an assemblage. One moves about in whatever assemblage they find themselves and diffractively works with the forces that shape their movements within the assemblage. In this way, the assemblage always positions one as a novice as the assemblage always contains surprising interactions that position one as a novice. Knowledge is not transmitted from stable subject positions of "novice" or "expert." We are all both novice and expert engaged in a series of becomings-with as we generate knowledge together. One is always in a state of becoming both expert and novice. Mentoring becomes an ongoing practice of ethics aimed at opening up becoming both expert and novice. The degrees of expert and novice generated in assemblages demand a different practice of mentoring – string figure mentoring.

String figure mentoring

Positioning qualitative inquiry as a territorial assemblage dismantles the binary division of expert/novice, the foundational binary of mentor/mentee. Consequently, a new term is warranted, what I am calling string figure mentoring. String figure mentoring is about creating and maintaining educative and generative relationships that have no objective definition. In other words, string figure mentoring is not something that *is*. Rather, it is an ongoing practice that creates moments of becoming-with each other.

While Haraway (2016) described string figures as multispecies stories, she briefly mentioned how string figures might work in scholarship. She wrote:

Playing games of string figures is about giving and receiving patterns, dropping threads and failing but sometimes finding something that

works, something consequential and maybe even beautiful, that
wasn't there before, of relaying connections that matter, of telling
stories in hand upon hand, digit upon digit, attachment site upon
attachment site, to craft conditions for finite flourishing on terra,
on earth. String figures require holding still in order to receive and
pass on. String figures can be played by many, on all sorts of limbs,
as long as the rhythm of accepting and giving is sustained. Scholarship and politics are like that too – passing on in twists and skeins
that require passion and action, holding still and moving, anchoring and launching.

(p. 10)

Haraway assumes that scholarship works within a flat ontology, one that
is characterized by exchanges, much like a territorial assemblage described
earlier. In such an ontology, scholarship is generated by movement, process, and forces. The forces create conditions of possibility that both reterritorialize (or anchor) and deterritorialize (or launch) scholarship. Specific
to mentoring, mentoring shifts to facilitate exchanges that both give and
receive. In this way, no one is solely an expert. No one is solely a novice.
Mentoring crafts conditions of possibility. Mentoring shifts to exchanges
that aim to generate something that perhaps could not have been thought
of or done before. These exchanges are constituted by the telling of stories, passing them from hand to hand. In this way, the exchanges require a
humble, still, and attuned listening. Mentoring takes on a practice of ethics
that is grounded in ongoing exchanges that seek to bring about something
new. String figure mentoring, then, is a verb, an ongoing ethical practice of
generative exchanges within a territorial assemblage that cannot be defined
in advance of the exchanges.

Deleuze's (1969/1990) events are useful to think about these generative
exchanges. In string figure mentoring, the people involved are always asked
to "become worthy of what happens to us" (p. 149). For example, I have
mentored for the American Educational Research Association Qualitative
Research SIG for some time now. Many times, I enter into those mentoring
relationships without much prior knowledge about the person. In this way,
I have to become worthy of what happens when those relationships materialize through events. In so doing, those mentoring relationships "will and release

the event" (p. 149) that constitute the relationship. As that event is released, the mentee and I become the offspring, iterations of both expert and novice, of those events (Deleuze, 1969/1990). The event, then, is a rebirth of those subject positions in string figure mentoring. String figure mentoring, then, is a series of events that creates offsprings of both expert and novice storytellers.

These series of events demand a different kind of ethics. String figure mentoring asks us "not to be unworthy of what happens to us" (Deleuze, 1969/1990, p. 149). As string figure mentoring events create expert and novice offspring, it is imperative to work against upholding hierarchical power relations that ultimately uphold static definitions of expert and novice. How might one open up becomings rather than foreclose them? How might one become worthy of becoming a different offspring of the event? Stengers's (2010/2011) work with cosmopolitics is useful in considering these questions. She wrote:

> The experience, always in the present, of the one into whom the other's dreams, doubts, hopes, and fears pass. It is a form of asymmetrical reciprocal capture that guarantees nothing, authorizes nothing, and cannot be stabilized by any constraint, but through which the two poles of the exchange undergo a transformation that cannot be appropriated by any objective definition.
>
> (p. 372)

String figure mentoring events are those dreams, doubts, hopes, and fears, passing through people. As those events pass through bodies, they undergo a grasping that creates and transforms people. Consequently, there can be no stable definition of string figure mentoring. Rather, it is a process of transformations happening within assemblages.

These transformations must work against the "re-circulating domination" (Trinh, 1991, p. 15) as they create possibilities. While the student in the extended example and I have differing identities (e.g., gender and [dis]ability), we both share a White Western cultural heritage that allows for exchanges between expert and novice in which novices can teach experts. Other people I have worked with have different cultural ideas about experts and novices that are critical parts of those transformations. In order to respect and honor the cultural heritage and practices of the people

I work with, I have to respect and honor how they may understand and practice relationships between experts and novices. Such work resists the re-circulation of dominance (Trinh, 1991), particularly the dominance of White Western ways of thinking and being that dominate most academic settings. A critical move of string figure mentoring is the active resistance of re-circulating dominant ways of knowing and being so as to create a plurality of voices in the qualitative inquiry assemblage.

The transformations of string figure mentoring are practices of extending hands to a qualitative inquiry assemblage. However, these practices are far from linear. The to-and-fro movements of events that create becomings disrupt even what work is. Trinh (1991) wrote, "every step taken is at once the first (a step back) and the last step (a step forward) – the only step, in a precise circumstance, at a precise moment of (one's) history" (p. 15). String figure mentoring recognizes that each step, each extension of hands happens at a particular time and context of one's life. Each event that creates different offspring in string figure mentoring, then, creates a moment in the qualitative work-in-progress. This work-in-progress, as Trinh wrote, "is not a work whose step precedes other steps in a trajectory that leads to the final work. It is not a work awaiting a better, more perfect stage of realization" (p. 15). In other words, string figure mentoring events are not linear and do not lead to a "finished product" that is a better, more perfect piece of scholarship. There can be no beginnings or endings in string figure mentoring. Nor can string figure mentoring produce perfection. Instead, string figure mentoring is about openings and closings. Trinh wrote that "closures need not close off; they can be doors opening onto other closures and functioning as ongoing passages to an elsewhere (-within-here)" (p. 15). In this way, string figure mentoring views closures as portals to other possibilities.

This is not to say, however, that openings in a qualitative project are in a space of pure possibility. As I mentioned earlier, each qualitative project is generated by the combinatory forces of a qualitative inquiry assemblage, academic discipline assemblage, and many times a departmental assemblage. These forces create the conditions, the laws and limits (Trinh, 1991), of possibility within a qualitative research work. As the closures and openings modify a work, each work becomes a work-in-progress, full of events yet to be actualized. String figure mentoring gets to work in these openings and closings of a work-in-progress. String figure mentoring, then, creates conditions for qualitative projects that work against the neoliberal norms of publish or perish in which closures serve "to wrap up a product and facilitate consumption" (Trinh, 1991, p. 16). String figure mentoring works against the ways in which the neoliberal academy stratifies bodies and inquiry into conformity, placing qualitative inquiry into slots that offer little to no freedom of movement. String figure mentoring aims to create tranquil and fertile spaces for works-in-progress to flourish.

String figure mentoring is about these passes that change us, that craft different avenues of becoming. In this way, string figure mentoring seeks to practice what Deleuze (1968/1994) said about education. He wrote:

> We learn nothing from those who say: "Do as I do." Our only teachers are those who tell us to "do with me," and are able to emit signs to be developed in heterogeneity rather than propose gestures for us to reproduce.
>
> (p. 23)

In other words, string figure mentoring consists of events that multiply heterogeneity rather than moments that seek to create reproductions of the same. It is about listening and caring for each other, creating moments of passage on the qualitative inquiry assemblage so that something different can be done, something that is unique to that person. As we do qualitative inquiry with each other in assemblages, we pass each other patterns to create possibilities rather than foreclose them. This is string figure mentoring.

As I finish this chapter, a flurry of string figure mentoring events is transforming me and others due to COVID-19. String figure mentoring in a pandemic is a revved-up version of pre-pandemic string figure mentoring. Daily life, scholarship, and other assemblages were always already entangled in pre-pandemic string figure mentoring. Pandemic string figure mentoring has amplified these entanglements. Pre-pandemic events always passed through us without warning. Pandemic events have amplified this lack of anticipation as new normals seem to be created daily. Pre-pandemic events asked us to be worthy of them. Pandemic events have made this sometimes painfully apparent and crucial.

No longer do mentees and I sit in my office, occasionally enjoying a lemon drop candy as we pass dreams and concerns about a project with each other. Now, if the weather is nice, I walk in my backyard as I chat with people on the phone. I answer emails. I video conference. Through whatever means available, we have been passing our dreams, our hopes, our fears, our doubts, sharing them so we can become transformed by them (Stengers, 2010/2011). Each conversation begins by passing COVID-19 events that are transforming our everyday lives, sometimes through direct connections to COVID-19, which has, unfortunately, included deaths of family members and friends. Sharing these events allows us to learn from each other and in turn become transformed by them, in effect learning how to live together in a pandemic. As we share these events, we learn about each other's lives – our families, jobs, and so on. Then, if appropriate (because sometimes the life events are just too much), we pass scholarship dreams, hopes, fears, and doubts. Many times, if not all of them, these passes are deeply entangled with everyday life events. These passes dismantle any existing binary of mentor/mentee and expert/novice as each and every one of us is learning how to live, both personally and professionally, again. For example, many times the person I am talking to shares that they find themselves unable to focus on their research. I frequently share that I, too, am having difficulties focusing on scholarship (even on finishing this chapter). If the person shares that they find their research area pointless now, we talk about how everything has changed and that maybe a shift in our research areas is not such a bad thing. Perhaps this doubt is a way of responding to, of being in touch with, the liveliness of the world (Barad, 2012). It seems we are all learning to live again as these events pass through us and transform us into new offsprings, just learning how to walk again as humans and then again as scholars.

String figure mentoring: to pass and to craft

String figure mentoring disrupts a transmission model of mentoring, one in which a mentor transmits knowledge to a mentee. Such a model limits both a mentor and mentee and keeps them firm within their categories creating little room for professional, much less personal growth. Instead, string figure mentoring gets to work in the qualitative inquiry territorial assemblage that "continually passes into other assemblages" (Deleuze & Guattari, 1980/1987, p. 325). In so doing, the assemblage transforms itself by generating new iterations of itself. These iterations are works-in-progress as they move between spurts and stops (Trinh, 1991). As these assemblages pass through each, they create events that can never be anticipated. Binary divisions such as mentor/mentee and expert/novice dissolve in these events as the events create both-and offspring. String figure mentoring works in the midst of these passages and aims toward making the multiple of qualitative inquiry. To do this work, string figure mentoring takes to task the infinitive verbs "to pass" and "to craft."

The verbs "pass" and "craft" figure heavily into Haraway's (2016) definitional work on string figures. Likewise, string figure mentoring takes on the importance of these verbs in its definition and complicates them by placing them in the infinitive form. Palmer (2014) wrote that "the infinitive is the undetermined problem and the conjugation is the solution" (p. 87). In other words, the infinitive verb creates the conditions to think of a problem as an entanglement of forces generated by assemblages passing through each other that have no answers, only possibilities (Deleuze, 1986/2006). The infinitives "to pass" and "to craft" then carry with them an imperative to work with the events of those shifting assemblages. This working with, thinking with shifting assemblages and the events they generate, become works-in-progress that have openings and closings (Trinh, 1991).

Conjugations of working within moving assemblages, and the forces these conjugations generate, create transformative becomings-with. The "to pass" of string mentoring consists of the passing of stories that materialize the forces at work in a project. This work is achieved by the stories we tell about our projects, how they are transforming and how we may be transformed by them in assemblages that entangle us. These transformations engage the verb "to craft." "To craft" is the work of conjoining those forces to create works-in-process that transform the qualitative inquiry territorial assemblage. These infinitive verbs, to pass and to craft, engage politics and ethics. Events materialize political forces of assemblages and demand an ethics that cannot be anticipated in advance. These ethical moments ask us to be worthy of them (Deleuze, 1969/1990) and ask us to foster transformative becomings-with.

Haraway's (2016) thinking about stories is instructive. She wrote, "Each time a story helps me remember what I thought I knew, or introduces me to new knowledge, a muscle critical for caring about flourishing gets some aerobic exercise. Such exercise enhances collective thinking and movement too" (pp. 115–116). While my main example in this chapter focuses on a student who did post-qualitative work, my specialty area, it is not reflective of all the mentoring I do both within and outside of my institution. From Black Feminism, Critical Race Theory, Queer theories, Postcolonialism, and others, it is my job to engage paradigmatic proliferation as a fertile ground for creating more lines in the qualitative inquiry assemblage. String figure mentoring engages epistemological and ontological difference rather than reducing to the same. This work asks qualitative research mentors to honor all paradigms, all ways of knowing and being, and foster becomings-with within those ways of knowing and being. For me, the work of string figure mentoring is creating space for students to contribute their own lines, crafted by the assemblages that entangle them – their personal desires, political settings, and . . . and . . . and . . . – that craft possibilities. String figure mentoring is exciting work, the kind of work that allows more people to find themselves entangled in the qualitative inquiry assemblage and create lines of what is yet to come in qualitative inquiry. String figure mentoring is about animating possibilities for all, proliferating difference, and honoring the vast ways of knowing and being that exist and are to come.

String figure mentoring is an additive practice that seeks to enlarge the world of mentoring in qualitative inquiry (Haraway, 2016). It is not a rejection of, a response to, or an opposition to other ways of thinking and doing mentoring. Like the authors cited in this chapter (Buck et al., 2009; Doloriert et al., 2012; Engels-Schwarzpaul & Peters, 2013; Gordon, 2003; Knox et al., 2011), string figure mentoring came from an intense thinking with both theory and practice that is based on my personal experiences, the political settings that I occupy, and the theories I think with. String figure mentoring is not better; it is just a different way of thinking and doing mentoring that works for me and perhaps for the reader. Moreover, the work done in this chapter does not reject Denzin's (2008) big tent metaphor. Instead, this chapter adds a different way to think about qualitative inquiry as an assemblage that helps me think about the mentoring work I do within it. String figure mentoring adds to Denzin's (2008) theses about qualitative inquiry by amplifying the call for diverse paradigms in qualitative research and honors the personal and political ways we come to these paradigms by offering a practical way of mentoring in qualitative research. In effect, string figure mentoring practices what it preaches. It does not seek to reduce. It seeks to enlarge by responsively and respectfully honoring the assemblages that generate it.

String figure mentoring is what I think and hope I do when I mentor someone. In these shifting assemblages, the person and I think and work, pass and craft, together through entanglements. "To pass" and "to craft" are very much similar to what Barad (2012) calls being in touch with the world. "To pass" and "to craft" are "never pure or innocent. [They] are inseparable from the field of differential relations that constitute [them]" (p. 215). In so doing, "to pass" and "to craft" are political and ethical movements of being in touch with the liveliness of those relations that create different configurations of the world (Barad, 2012). These passings and craftings in entanglements transform us into a both-and of expert and novice and, in turn, transform the assemblages in which we work, think, and become something otherwise. String figure mentoring – to pass, to craft – new becomings within assemblages that transform us and the world.

References

Barad, K. (2012). On touching – the inhuman that therefore I am. *Differences: A Journal of Feminist Cultural Studies, 23*(5), 206–223. https://doi.org/10.1215/10407391-1892943

Bergson, H. (2002). The possible and the real. In K. A. Pearson & J. Mullarkey (Eds.), *Henri Bergson: Key writings* (pp. 223–232). Continuum. (Original work published 1920)

Buck, G. A., Mast, C. M., Macintyre Latta, M. A., & Kaftan, J. M. (2009). Fostering a theoretical and practical understanding of teaching as a relational process: A feminist participatory study of mentoring a doctoral student. *Educational Action Research, 17*(4), 505–521. https://doi.org/10.1080/09650790903309375

Constantino, C., Manning, W., & Nordstrom, S. (2017). Rethinking covert stuttering. *Journal of Fluency Disorders, 53*, 26–40. https://doi.org/10.1016/j.jfludis.2017.06.001

Deleuze, G. (1990). *The logic of sense* (M. Lester with C. Stivale, Trans. & C. V. Boundas, Ed.). Columbia University Press. (Original work published 1969)

Deleuze, G. (1994). *Difference and repetition* (P. Patton, Trans.). Columbia University Press. (Original work published 1968)

Deleuze, G. (2006). The brain is the screen. In A. Hodges & M. Taormina (Trans.) & D. Lapoujade (Ed.), *Two regimes of madness: Texts and interviews, 1975–1995* (pp. 282–291). Semiotext(e). (Original work published 1986)

Deleuze, G., & Guattari, F. (1987). *A thousand plateaus: Capitalism and schizophrenia* (B. Massumi, Trans.). University of Minnesota Press. (Original work published 1980)

Denzin, N. K. (2008). The new paradigm dialogs and qualitative inquiry. *International Journal of Qualitative Studies in Education, 21*(4), 315–325. https://doi.org/10.1080/09518390802136995

Doloriert, C., Sambrook, S., & Stewart, J. (2012). Power and emotion in doctoral supervision: Implications for HRD. *European Journal of Training and Development, 36*(7), 732–750. https://doi.org/10.1108/03090591211255566

Engels-Schwarzpaul, A. C., & Peters, M. A. (Eds.). (2013). *Of other thoughts: Non-traditional ways to the doctorate: A guidebook for candidates and supervisors*. Sense Publishers. https://doi.org/10.1007/978-94-6209-317-1

Foucault, M. (1978). *The history of sexuality: Volume 1, an introduction* (R. Hurley, Trans.). Vintage Books. (Original work published 1976)

Foucault, M. (1985). *The use of pleasure: Volume 2 of the history of sexuality* (R. Hurley, Trans.). Vintage Books. (Original work published 1984)

Foucault, M. (1986). *Care of the self: Volume 3 of the history of sexuality* (R. Hurley, Trans.). Vintage Books. (Original work published 1984)

Gordon, P. J. (2003). Advising to avoid or to cope with dissertation hang-ups. *Academy of Management Learning & Education, 2*(2), 181–187. www.jstor.org/stable/40214183

Haraway, D. J. (2016). *Staying with the trouble: Making kin in the Chthulucene*. Duke University Press. https://doi.org/10.1215/9780822373780

Haraway, D. J. (2018). SF with Stengers: Asked for or not, the pattern is now in your hands. *SubStance, 47*(1), 60–63. www.muse.jhu.edu/article/689014

Jackson, A. Y., & Mazzei, L. A. (2012). *Thinking with theory in qualitative research: Viewing data across multiple perspectives*. Routledge. https://doi.org/10.4324/9780203148037

Knox, S., Burkard, A. W., Janecek, J., Pruitt, N. T., Fuller, S. L., & Hill, C. E. (2011). Positive and problematic dissertation experiences: The faculty perspective. *Counseling Psychology Quarterly, 24*(1), 55–69. https://doi.org/10.1080/09515070.2011.559796

Lakoff, G., & Johnson, M. (2003). *Metaphors we live by*. University of Chicago Press. https://doi.org/10.7208/chicago/9780226470993.001.0001

Palmer, H. (2014). *Deleuze and futurism: A manifesto for nonsense*. Bloomsbury Academic. https://doi.org/10.5040/9781472594402

Roulston, K., & Bhattacharya, K. (2018). Teaching in the era of the big tent: Presenting proliferation and polyphony. *International Review of Qualitative Research, 11*(3), 251–255. https://doi.org/10.1525/irqr.2018.11.3.251

Stengers, I. (2011). *Cosmopolitics II* (R. Bononno, Trans.). University of Minnesota Press. (Original work published 2010)

Tracy, S. J. (2010). Qualitative quality: Eight "big-tent" criteria for excellent qualitative research. *Qualitative Inquiry, 16*(10), 837–851. https://doi.org/10.1177/1077800410383121

Trinh, M. H. (1991). *When the moon waxes red: Representation, gender, and cultural politics*. Routledge. https://doi.org/10.4324/9780203700624

Watkins, M. (2010). Desiring recognition, accumulating affect. In M. Gregg & G. J. Seigworth (Eds.), *The affect theory reader* (pp. 269–285). Duke University Press. https://doi.org/10.1215/9780822393047

MENTORING AS A COLLECTIVE RELATIONALITY

Nikki Fairchild

Mentoring suggests didactic knowledge transmission, the mentor as trusted advisor training junior colleagues. There are alternative, more open and connected conceptions of mentoring practices (Nahmad-Williams & Taylor, 2015). This "mentoring moment" builds on work inspired by post-humanist and new material feminist inquiry (Taylor & Hughes, 2016; Koro-Ljungberg, 2016), proposing mentoring in undisciplined qualitative research as a collective relationality.

> *Mentoring is not the right word;*
> Choreography of bodies;
> Research-creation;
> More-than-human relationality;
> Response-able practices;
> Sensorium;
> Thinking-doing.

Research-creation-workshop-experiment-events are assemblages of relationality that activate undisciplined methodological interventions disrupting traditional research methods (Manning & Massumi, 2014). Experimental methods open alternative ethico-onto-epistemological spaces

DOI: 10.4324/9781003022558-25

for more inclusive, diverse, relational, and affirmative methodological approaches to "method," "knowledge," and "ways of knowing" (see Benozzo et al., 2019).

> Undisciplining of (research)practices;
> Connection – people, objects, things;
> Nourishment – physically, philosophically, materially.

Acknowledging bodies that are unnoticed allows for an unsettling of the human "as the historical site of political privilege to include a broader range of ontologically diverse actors" (Taylor & Fairchild, 2020, p. 513). Bodies involved in research-creation include humans, nonhumans and other-than-humans; all have mentoring potential as part of co-relational constellations of assemblages. These connections produce flows, forces, and intensities through which "life" is materialized, helping us rethink what constitutes a mentoring "experience."

Mentoring as a collective relationality is response-able to affective resonances between bodies. Mentoring, therefore, is not a direct transmission of knowledge; it is immersive, material, multi-modal, a sensorium of thinking--doing practices where "learning" and "support" are relational to human, nonhuman, and other-than-human connections. I am part of an international group of researchers working together since 2016 on experimental research-creation practices that seek to undiscipline qualitative research. We are connected by more than our personal friendships; we are connected relationally in spacetimemattering by the material conference events and the assemblages that form before, during, and after the events are choreographed. Barad (2007, p. 49) proposes that "knowing does not come from standing at a distance . . . but rather from a direct material engagement with the world." Mentoring as a collective relationality acknowledges this material engagement and how more-than-human subjectivities produce nourishing relationships between bodies, enhancing how undisciplined qualitative research is enacted.

Image 9.1 Mentoring as a collective relationality

Source: Nikki Fairchild

References

Barad, K. (2007). *Meeting the universe halfway: Quantum physics and the entanglement of matter and meaning.* Duke University Press.

Benozzo, A., Carey, N., Cozza, M., Elmenhorst, C., Fairchild, N., Koro-Ljungberg, M., & Taylor, C. A. (2019). Disturbing the AcademicConferenceMachine: Post-qualitative returnings. *Gender, Work and Organization, 26*(2), 87–106.

Koro-Ljungberg, M. (2016). *Reconceptualizing qualitative research: Methodologies without methodology.* Sage.

Manning, E., & Massumi, B. (2014). *Thought in the act: Passages of ecology of experience.* Minnesota University Press.

Nahmad-Williams, L., & Taylor, C. A. (2015). Experimenting with dialogic mentoring: A new model. *International Journal of Mentoring and Coaching in Education, 4*(3), 184–199.

Taylor, C. A., & Fairchild, N. (2020). Towards a posthumanist institutional ethnography: Viscous matterings and gendered bodies. *Ethnography and Education, 15*(4), 509–527.

Taylor, C. A., & Hughes, C. (2016). *Posthuman research practices in education.* Palgrave Macmillan.

10

MENTORING (MAYBE) AS A PHILOSOPHICAL EVENT

Mirka Koro, Mariia Vitrukh, Nicole Bowers, Lauren Mark, and Timothy C. Wells

What if *mentoring* functioned as an *invitation* and *opening* to *relate, differently*? Perhaps then mentoring programs and institutional mentoring agreements could serve as productive beginnings, finding potential through processes of event-ing.

Certainly, mentoring both inside and outside of academia has interested many scholars. Concerns over rates of student attrition and mental health disorders, both too high, have led researchers to study mentor-mentee relationships and find that they have a profound effect on graduate student well-being, retention, and academic success (Al Makhamreh & Stockley, 2019; Council of Graduate Schools, 2010; Roberts et al., 2019). Many maintain that good mentorship *encourages* both the personal and professional growth of students, suggesting that the work of a mentor extends beyond academic competencies and into the realm of psychosocial, affective competencies (Al Makhamreh & Stockley, 2019; Brondyk & Searby, 2013; Roberts et al., 2019; Yob & Crawford, 2012). Although some see supervising, advising, and mentoring as interchangeable, most studies distinguish mentoring for its focus on personal development aspects of relationships (Brondyk & Searby, 2013; Woloshyn et al., 2019) and claim that mentors can help their

DOI: 10.4324/9781003022558-26

211

students overcome personal challenges (Brondyk & Searby, 2013; Roberts et al., 2019). Moreover, researchers find that mentoring can enhance the success of traditionally marginalized students (Chan, 2008; Dedrick & Watson, 2002; Noy & Ray, 2012; Rose, 2005) particularly, women (Mansfield et al., 2010), international students (Wang & Li, 2011), and African American students (Felder, 2010; Grant & Ghee, 2015). All of this mentoring literature suggests that mentors play a vital role that stretches well beyond their own academic training.

Although programs and educators have explored distributed models of formal doctoral mentoring (Kram, 1988; Mullen et al., 2010), the apprentice model of individual mentoring continues to be the dominant approach to mentoring (Hall & Burns, 2009). At its most basic, mentorship is an exchange of both knowledge and experience (Lyons et al., 1990). A more specific understanding acknowledges the practical and professional guidance that the mentor provides to the mentee (Samier & Fraser, 2000). Yet the nature and boundaries of this relationship is rarely as simple as an **e x c h a n g e**. For many, understandings of mentorship require the consideration of the social and community practices that sustain the relationship (Jacobi, 1991). Although emotional and academic support are offered through mentoring, mentoring often remains a unidirectional experience with a somewhat hierarchical structure where the mentor takes on the bulk of the responsibility for shepherding the mentee and the mentee absorbs the mentor's expertise and support in rather passively modeling approaches to mentoring (e.g. Blackwell, 1989). We are interested in further exploring whether mentorship is a hierarchical structure including voluntary or directed relationships, whether it occurs within or outside of professional settings, and whether or not it involves formal evaluations (e.g. Arnesson & Albinsson, 2017), all within the context of qualitative inquiry.

It can be challenging for institutionalized, pre-planned, and highly formalized mentoring to adapt to changing circumstances and shifting social contexts. To emphasize creative relationality and immanent questions of knowledge within mentoring processes in academic and qualitative inquiry contexts, this chapter positions mentoring as a series of relational events which are always philosophical yet driven by practices associated, and not so closely associated, with qualitative inquiry. These philosophical and relational events can be initiated by thought and care, perhaps by qualitative inquiry traditions, but they do not know how to speak without hesitation or how to manifest themselves orderly and logically in *different time-spaces*. Qualitative inquiring time-spaces can be planned, scheduled, and pre-organized, but these events can also be emergent and continuously shape how time and space are lived and experienced during "mentoring" and inquiry processes. Mentoring as a series of philosophical events might also function through active creation yet uneasy relationality, thus also shaping collective "subjects" of qualitative research. If for Deleuze and Guattari (1994), philosophy is about creating concepts, similarly, mentoring as a philosophical event calls for immanent creation (of reactions, responses, relations, relational selves, ways of togetherness, co-living and more). Rather than setting up an a priori planned relationship, mentoring as a philosophical event is relational, experimental, and returns without promises. In this chapter we draw from examples of mentoring in higher education, specifically from PhD programs, and from qualitative inquiry to illustrate how mentoring may function as a condition for possibility and a living practice that is (self-)posing and (self-)creating while concerned with multiplicity, diversity, and various affects. Mentoring as a condition for possibility does not guarantee "mentoring," but it allows different "mentoring" practices to emerge and evolve. Nothing *must* involve mentoring, but everything potentially *could*. Thinking about mentoring in qualitative inquiry in this way places the responsibility and creation of mentoring within the everyday actions and choices of scholars, teachers, and students. It does not dictate what mentoring must look and feel like, but it offers a framework where mentoring in the context of qualitative inquiry can continuously become (different) and be practiced differently. The following relational events do not form an empathic image of mentoring or represent holistic mentoring practices, but together they compose unfinished sketches of diverse ways of relating.

In this chapter, we use philosophy as a practice and strategy to process mentoring flows, relational moments, and our lives as mentors and mentees (always both and more than one). While we sense and acknowledge more and less direct past and presence of philosophy in our activities and inquiry processes, we cannot directly and neatly identify, point to, or describe where, how, or when philosophy flows and interacts with our (and other's) lives, qualitative inquiry practices, and sensed mentoring events. Almost nothing is uniquely personal and subjective (and thus our collective "we" and multi-author) in this text. Yet these events, similar to all events, are still contextual and simultaneously affective. Finally, much of our *sensing* and *wondering* in this chapter could be seen as non sensical and we would like to keep it that way. We propose that mentoring does not need to mean anything, and we do not need to understand its processes. Rather, the overanalysis, overly descriptive inquiry processes and uses of methodological strategies, humanistic mentoring agendas, and cognitive and psychological mentoring reflections, are likely to limit experiences and mentoring's different becomings.

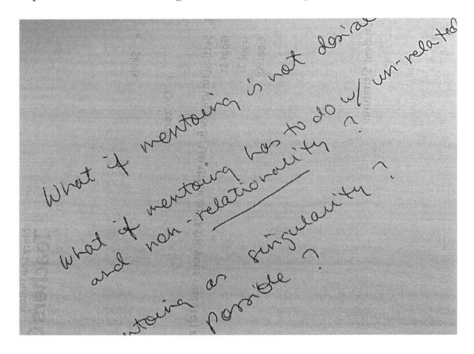

What if mentoring is not desirable? **Event 2**

What if mentoring has to do with un-relatedness and non-relationality? What if mentoring functions as singularity?

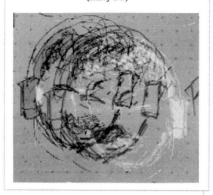

(maybe)

In some contexts, academic and professional success of students can be defined through publications and degree completion (Brill, Balcanoff, Land, Gogarty, & Turner, 2014; Roberts, Tinari, & Bandlow, 2019; Yob & Crawford, 2012). Literature states that collaboration and co-authoring might also help students to develop the critical areas of scholarly writing and employment of research methods, but most graduate students and faculty mentors believe that this also entailed more than just competence and expertise on the side of the mentor (Roberts, Tinari, & Bandlow, 2019; Yob & Crawford, 2012).

Studies also indicate that an effective mentor is encouraging, respectful of student ideas and direction, and is both physically and emotionally available to students (Al Makhamreh & Stockley, 2019; Roberts, Tinari, & Bandlow, 2019; Woloshyn et al., 2019; Yob &Crawford, 2012). According to literature, it can also be important to support the well-being of students through a variety of strategies ranging from simple encouragement to adopting attitudes of nurturing and caring (see e.g., Al Makhamreh & Stockley, 2019; Roberts, Tinari, & Bandlow, 2019; Woloshyn et al., 2019).

Philosophies of mentorship hinge not only on understandings of the mentor-mentee relationship but also on understandings of learning as such. Much of the research theorizing mentoring relationships draws social processes theories of learning suggesting that the relationship is one of expert and novice (Kay, Hagan, & Parker, 2009). The implication is not only that the "identity and status" of each mentor and mentee is firmly established but that the outcome of the relationship should be known in advanced, i.e. mutual expertise. Some have questioned the presumed directionality of the mentor relationship, opening the door towards artful experience which can take the relationship in new and creative directions (Flint & Guyotte, 2019). Others have similarly revisited the outcome of the relationship, emphasizing the return of the mentee as a "superaddressee," something tertiary to the relationship (Bryzzheva, 2006).

in the context

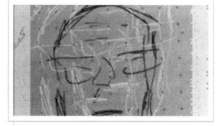

Mentoring can be one of the most important tasks of academics, teachers, and educators; it is a way of collective living, aspects of responsible and ethical scholarship, and one important way to give back to (learning) communities. More specifically, our mentoring activities vary across time and space since we engage in mentoring which is driven by mentor-mentees' situational and specific needs. Oftentimes mentor-mentees intensively

exchange

transformative dialogue

Koro-Ljungberg and Hayes (2006), approach mentoring from a relational stance proposing that "when mentors and mentees enter relationships, they bring with them multiple relational histories that represent a unique combination of communities and voices" (p. 393). Thus, a list of best practices may not provide the best path toward mentoring, and they suggest that mentors and mentees engage in transformative dialogue that makes "their own epistemological, theoretical, and practical interests transparent" (Koro-Ljungberg & Hayes, 2006, p. 404) so as to take joint responsibility.

Scholars studying mentoring ask questions about effectiveness, effective mentoring roles, and types of relationships (see e.g., Kraiger, Finkelstein, & Varghese, 2019). Approach to mentoring is usually shaped by "both how we were mentored and the hidden curriculum in higher education" (Nagasawa, & Swadener 2016, p. 184). Mentoring in qualitative research, as most academic graduate mentoring, involves a lasting relationship (often over several years) and high-stakes outcomes: successful completion of the dissertation, set of recommendations for employment, publication and conference opportunities, collaborative practices and networking (Duffy, Wickersham-Fish, Rademaker, & Wetzler, 2019).

Instead of looking at mentorship relationships from the perspective of doctoral students (Burkard, Knox, DeWalt, Fuller, Hill, & Schloesser, 2014; King & Williams, 2013; Koro-Ljungberg, & Hayes, 2006; Spaulding & Rockinson-Szapkiw, 2012) or faculty members (Knox, Burkard, Janecek, Pruitt, Fuller & Hill, 2011), mentoring relationships in qualitative research may be viewed as the apprenticeship model of graduate education where knowledge, skills and academic habitus are transferred through co-constitutive relationship or reciprocity (Nagasawa, & Swadener, 2016). The unexpectedness of such reciprocal relationships allows scholars to tap into the problems, dilemmas, opportunities, ethical issues and persistent power relations. Uncertainties of faculty life – "competition over increasingly scarce tenure track appointments, continual job searches, and navigating the universe of institutional prestige, all while teaching and remaining active as a scholar" (Nagasawa, & Swadener 2016, p. 173) adds another layer to the complexity of this relationship.

participate in shared academic lives and therefore have a chance to experience the full range of academic and professional activities. These vary from weekly individual and group meetings discussing qualitative inquiry and its processes; to teaching examples and activities related to qualitative inquiry classes; to late night emergency phone calls and text messages; to co-teaching, collaborative planning, shared conference presentations and papers, national and international networking, and collective international travels; to assisting students with their job application materials, job talks, postgraduate experiences, and more. Mentoring evolves and transforms throughout often intense relationships. Even though office hours and institutionalized mentoring activities and relationships can be productive and generate relationality, we have found that organic and situational mentoring can be, or maybe always is, as or more productive. Mentoring often happens outside the classrooms: at home, coffee shops and cafeterias; via Skype and phone; on conference trips, at schools, research sites and communities; while walking to the classroom across the campus; at new job sites; and, many other unexpected and emerging spaces. Or are we just imagining these potential mentoring events? Does the potential virtuality of mentoring take away from its impact?

(Pre-)event (to mentoring)

This chapter is not about Deleuzian philosophy per se; however, Deleuze's work created for us an opening to access mentoring events through ongoing creation and experimentation. We are certainly not the first set of authors to deploy Deleuze or event philosophy as a creative writing partner. In their collaborative writing project centering on Deleuzian concepts, Wyatt et al. (2011) draw on Deleuze's (1995) concept of the individual as a series of events, or interconnected currents passing through one, so that even the seemingly solitary act of writing alone in a room is recognizably conducted in the company of other influences. While describing their collaborative process, Wyatt et al. (2011) center Deleuze's concept of nomadic inquiry, specifying the rhizomatic nature of nomadic space. Viewing the process of writing itself through a nomadic lens, they characterize nomadic writing as a "shifting, creative process of conceptualisation" via "lines of flight taking off from the sedimented order of the already known, in new directions, plotting different courses, and engaging in experimentation" (p. 13). This type of rhizomatic space invites multiple entryways for authors working together in partnership. The distinction between "nomad space rather than . . . the nomad itself" (p. 14) remains relevant to a great deal of collaborative writing, whether undertaken among scholars in physical proximity to one another or connecting time zones and internet cloud storage. For us, the space of a blank document on Google Drive stretching in between individually penned events served as nomadic space. As social distancing rules and self-quarantining practices due to the novel coronavirus took effect midway through our process, our transition from in-person meetings to virtual check-ins served to further emphasize the role of nomadic space – the sole medium left for connection among authors geographically only a short drive apart.

Secondly, Wyatt et al. (2011) touch on Deleuze and Guattari's (1983) BwO, or Body without Organs, as a means for the authors to initiate "the idea of ourselves in relation, ourselves as co-extensive and co-implicated with the other" (p. 17). **All of us, as involved authors, have already been co-implicated with one another through the process of doctoral studies mentoring, although our mentoring experiences also extend beyond the individuals involved in this project.**

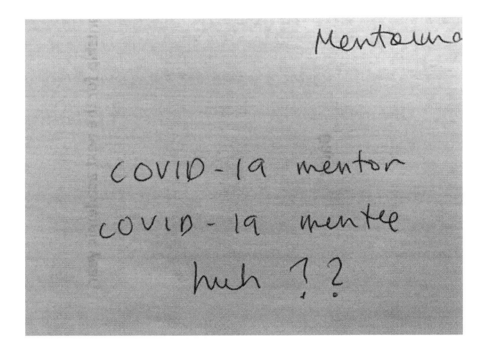

During our writing process, we made use of the comment function in Google Docs to continually implicate ourselves across all parts of the text. We did not use this as a means of revision or editing suggestions for each other's work, but as inventive suggestions that might give rise to further directions of unidirectional growth. We kept ourselves open to lines of flight that might be unexpected or divergent. Similarly to Wyatt et al.'s (2011) **spirit of collaboration**:

> we were not here to judge each other against some abstract stand-
> ard, nor here to insist that we follow some preconceived plan, nor
> yet were we to abandon affect, our bodily specificities, or the com-
> plex challenges that life threw up for each of us.
>
> (p. 134)

We simply invited linguistic wings of fancy guided by the words "what if?" to carry each event through further lines of flight.

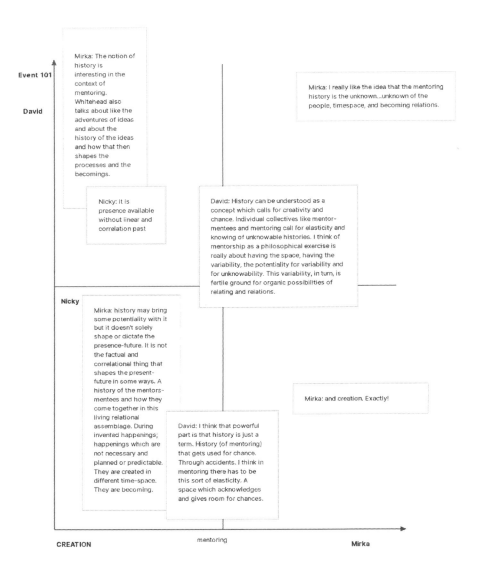

Sub-routine 102048

Another mentoring meeting.

Worry. Doubt. Hope. Excitement . What will our bodies in one space bring? What if we are behind, adrift, askew? What can we contribute to this relationship that is worth our time?

Time more precious than anything:

Monday at 11:00 am?

That won't work for me, I have another meeting. How about Tuesday at that time?

Sorry, I have class then. I am available Mondays 11–3, Tuesdays after 2, Wednesdays 9–11 and 2–6, Thursdays and Fridays anytime.

This week is pretty full, let's try for Thursday at 1 pm, but I will only have 45 minutes.
Okay, see you then!

Why do mentoring-time and time-mentoring seem to be the (only) force?

Preparation. Anticipation. The pulling of future scholarship, the drag of past scholarship, the chaos of present scholarship. Exhausting. Exhilarating.

Hey, I have to cancel our meeting this week. Let's meet Monday at 11.
Okay, I can just keep working. See you Monday.

Why do relational-time and time-relations seem to be the (absent) force?

Relief? Hurt? Determination? Swimming in the open ocean with no guidepost in sight. Should we tread for a while? **Float?** Another stroke burns with agony. Let's dive under, a bit deeper until we are swallowed whole.
Preparation. Anticipation. The pulling of future scholarship, the drag of past scholarship, the chaos of present scholarship. **Exhausting. Exhilarating.**
When we step through the threshold of the office, we potentially tug on our rather diffuse body, pulling ourselves together so they say. **Smile, sit, take out notebooks, pop open laptops, don't look crazy.**

How are you?
Great. *If by great I might mean I have been wrestling with the unfathomable chaos of the Universe, the cacophony is giving me a terrible headache, I have to take pills to sleep, and I am a raw feeling all of the time. Yeah, maybe not how we should start this meeting.* How are you?
Busy. We just submitted a grant, so I think some of my time will free up. So where are we at?
I started a draft of the methods section. *I wrote three different versions of the methods section, put them aside and wrote another. I have a whole two paragraphs of work to show, show that I did something other than take detours.* It is in the shared drive. I also got to a part in the

analysis that could use some talking through. *Like taking a drive and finding the road drops off into another dimension of space/time.* If you look in our drive, I put parts of the transcripts that I am not sure about. At the top, I wrote our research questions, and under that I outlined our theory, just to remind us of our direction. *Organized, this instills confidence, right? I am part of this creation, right? Right — feels like a tightrope walk over a pit of piranhas.*

> *Why do economical-time and time-economy*
> *seem to be the (motivating) force?*

Space/time coalesces, tempo picks up. **We work. We create.** We are unsure of where we end and they begin. Nirvana, enlightenment, wonder, awe. **Body buzzing, brimming, tingling with** the rush of creation, the atomized pieces of us. Words on a page, thoughts in the air, a cloud so dense that it holds me up even as we walk out the doors, taking them with us. Light with hope, supported by the intensity of 60 minutes of blissful generativity. We look back at our space, lean back through the door.

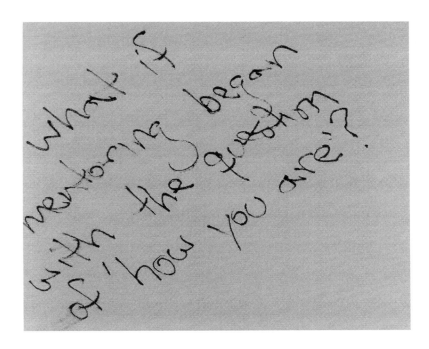

Thanks, that was fun. I feel great about this direction. We saved me again. How do you do that, how do you navigate the depths of you/me? I love you.

Why do care-creation-time and time-care-creation seem to be the (love) force?

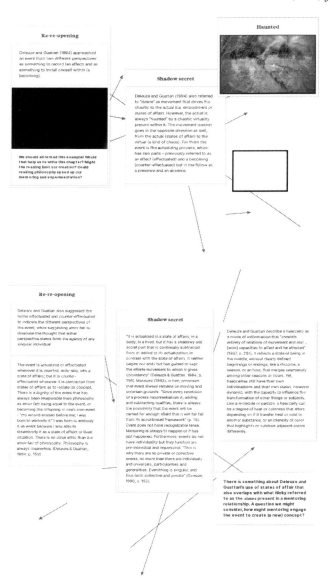

Event 14

What does mentoring as a philosophical encounter do?

It may enable mentor-mentees to ask difficult yet relational questions about truth, knowledge, presence, relationality, ethics, and materiality of our lives as scholars. It may bring philosophers (and their texts) into a dialogue with mentor-mentee. This philosophical dialogue, in turn, may extend our thinking-doing. If philosophy is about creating concepts, then philosophical mentoring could be about creating concepts of care, relating, collective difference, and meeting the other.

Me-**tour**
Mend-or
Mento-**ring**
Wasted-me-n-**to**-ring

Event 190.35Bgr/354

Okay. Something in this (perhaps the writing about well-being) reminded us about Shaviro's (2014) writing on Whitehead. He is talking about situations and antinomies and how two seemingly opposite entities can relate. They do things, inhibit, contrast, block, annihilate, merge, repel, create spacetime crevices. When one inhibits it works in opposition to the other, a kind of destructive force. When one contrasts, it forms a kind of pattern and reaches towards what Whitehead calls aesthetics or beauty. However, he also writes on the necessity of destruction for creativity: "Nonetheless, without such processes of destructive appropriation, there would be no self-enjoyment and no 'creative advance'" (Shaviro, 2014, p. 18). We wonder how we might think about mentoring in this inhibiting,contrasting, destroying relationship?

As with any event, mentoring may occur in the intense and violent collision of will have existed-ing universes temporary (temporally-spatially?) manifest as a mentor-mentee. The bodies of the mentor-mentee while seemingly individual seethe with innumerable inherited, simultaneous, and anticipated virtualities spread over a fuzzy space/time mattering. Whitehead (1967) explains that "[o]ur consciousness does not initiate our modes of functioning. We awake to find ourselves engaged in process, immersed in satisfactions and dissatisfactions, and actively modifying either by intensification, or by attenuation, or by the introduction of the novel" (p. 46).

We can imagine the meeting of such bodies, and in fact have been/are/will be such bodies, awakening to find themselves part of the mentoring event, entangled now in a new mentor-mentee body to varying degrees of self-enjoyment and concern. The patterns of intensity and contrast that occur here can become beautiful or not.

For Whitehead (1967) there exists various forms of beauty. At the most non-generative level we find the placid relationships of symmetry, harmony, overall likeness that some might term perfection. The affective forces here manifest as feelings of well-being, contentment, and ease. A mentor-mentee event that persists in this form of beauty may be full of enjoyment, but Whitehead would consider it to have less potential for creativity and spontaneity of ideas.

Perhaps an imagining of this type of mentor-mentee event would be fruitful. The individual bodies awaken in the stream of process of becoming. This "awakening" being the stage in the event of consciousness where we feel most familiar, but virtualities have been/are/will be in play that reside untouched by consciousness though forceful all the same.

expectations

These virtualities vary temporally, and it may be that the anticipated future remains less acknowledged than the inherited past and the simultaneous present. Within the mentor-mentee event Whitehead (1967) might say that the anticipated future has a great deal of pulling force on the present occasion. In other words, the expectations and plans the mentor/mentee hold, like publishing X number of papers, graduating and working at an R1 university, or fulfilling grant requirements exert a force that shapes the event.

Most of us consider those expectations to be speculative, but many may not consider the present and the past as equally speculative. What might happen if we treat all possible knowings as we do the anticipated future- all equally tentative and uncertain? Whitehead (1967)suggests that when we do that new ideas can manifest, ideas we could not have possibly imagined consciously. When we tread with certainty through time, we might experience the placid beauty and wellbeing mentioned above, but not much new will come about.

So, what might the alternative, Whitehead's (1967)Intensity Proper of Beauty, look/feel/smell/sound like? And, are we willing to suspend the conscious 'control' we feel with repetition, the certainty of a stable past, present, future? It might be important to consider that the 'control' we feel exists only loosely around events, and that the higher form of beauty arises in part from discord.

Event 2020

Here we are (again?).

    We have not been together for a week. Or are we always together? Are we in a mentoring relationship to infinity and beyond? Buzzz. . . .

    Something has changed? Or maybe it is just our attention drawn differently.

    How uncertain are we willing to be? Are we willing to go beyond (conscious) imagination? If we let our grasp open on what should be, could we allow us to expand in (im)possible directions? Let's try.

    The downturned corners of the mouth conflict with, "Alright, tell me more about that." The sigh and downcast eyes, conflict with, "Yeah, I can do that."

Conflict: When bodies use language to say one thing but viscerally express opposing sentiments. *Good, bad, creative? But that cannot be all. . . .*

    Here we are (again?).

    We met yesterday which was another day's tomorrow. We sit together in these familiar four walls, but we remain closed to spontaneous becoming with bodies directly at hand. Today, yesterday's tomorrow, we become not-together. Can we see the separate extra-spatial/temporal forces locked onto our becoming apart? What good is physical simultaneity if it cannot eclipse the unseen, the not physically present?

    "It would be a waste of your time," spoken with certainty and authority.

    Countered with, "It would be a valuable experience."

    We sit across from one another, mirrored expressions of hard eyes above grim thin lines of lips. Radiant with disappointment and disapproval. Paradoxically similar in bodily expression, we clash.

    Similar utterances and embodiment go on and on and on, looping with no outlet.

Clash: Presupposed solid bodies colliding unable to become together.

Does it have to be so stark?

This mentoring event could take infinite turns, trips, or tumbles. Will this event become out of **Authority** and **Power**? Through seeking **Easy, Harmonious** interactions? Through **Variance** and **Difference**? Will it fail to become a mentoring event at all? Or become in ways not-yet-imagined?

A quick trip through Google's dictionary beginning with the words **mentor** and **conflict** following related words brings us to fuzzily connected routes each and all as possible as any others found and not found. Tip, tap, typing words taking the first definitions- Google's algorithmic beats, dancing digitally. Defined? Finite? Embedded in strokes or taps the diffusion of expression and experience that weld the words Mentor and Conflict. Even within these paltry confines of lines constructed, construed, concatenated language, we experience the multiplicity seething under the cover of words that are themselves events.

Inhibition and/or **Annihilation**	A **Lowly** Form of Beauty	Beauty of **Intensity** Proper
Authority: *the power or right to give orders, make decisions, and enforce obedience* Similar: Power	**Harmonious:** *free from disagreement or dissent* Similar: Easy	**Variance:** *the fact or quality of being different, divergent, or inconsistent* Similar: Difference
Power: *the number of times a certain number is to be multiplied by itself* (mathematics)	**Easy:** *achieved without great effort; presenting few difficulties*	**Difference:** *a point or way in which people or things are not the same*
Destructive, damaging, a form of evil. **Paradoxically,** destructive forces also create. The more forceful the destructive nature the more chaotically creative when they implode with long ignored differences (Whitehead, 1967).	A symmetric, simple-minded, still beauty many call perfection in its placidness. Even and unchanging through lack of disruptive differences. **Boring** and unadventurous (Whitehead, 1967).	Difference and discord adapting, dancing together in eventfulness, stretched and strained but clearly woven into **patterns of contrast** that make an event an "intense experience" (Whitehead, 1967, as quoted in Shaviro, 2014, p. 42).

Although Shaviro (2014) argues this for Speculative Philosophy, we believe that Speculative Mentoring Events have "an irreducibly aesthetic dimension: [they] require new, bold invention rather than pacifying resolutions" (p. 43). The now and then can be conveyed through truly felt experience, but the future eventfulness must incorporate non-structural play with words as well. What words wield our futures as mentors and mentees? What fuzzy, diffuse collections of supposed meanings pull at our present bodies as they event? What expectations crouching in future mists manifest in today's tomorrows?

Speculative Qualitative Inquiry (SQI) runs the same rails. The future exerts pressure enough to shape nows as the past gushes over and around supplying surplus ingredients. This temporally immanent state leaves the inquirer with a ticker-tape parade of events that refuse to stand still, be captured, analyzed, and represented. Further complicating our matters, inquiry unfolds eventfully with(in) the events under inquiry. Knotted, not known, but all the more beautiful for the intensity of experience that is the act of inquiry open to Shaviro's (2014) "irreducibly aesthetic dimension" (p. 43).

One technique aligned with the intensity of speculative inquiry, is speculative writing for events of future reading through already, admittedly porous words. Language presents a challenge as traditionally it has been used to reduce experience to representation (Manning, 2013). Both Whitehead (1967) and Manning (2013) suggest that language morphs and blurs, remaining already/always inadequate to the task of actual representation. Far from being a disappointment, this leakiness allows speculative writers (and eventual readers) the opportunity to become with words and text. Manning (2013) asks us to attend to words differently explaining that, "[w]ords are never just what they seem to mean: they dance, they gallop, they tune in or out" (p. 164).

Words event with(in) events. Speculative writing retains affective tonality of experience, as words rupture out of the associated milieu, lured by a text's becomings. In playing with words, their position, their sounds, their rhythmic beats, their shapes, we open up text to more-than representation. In following words' loose association with one another, writers can become more-than-one during an intimate experience with words. In experiencing textual becoming, we must acknowledge that no communication comes fully formed no matter how linear it may first appear (Manning, 2013).

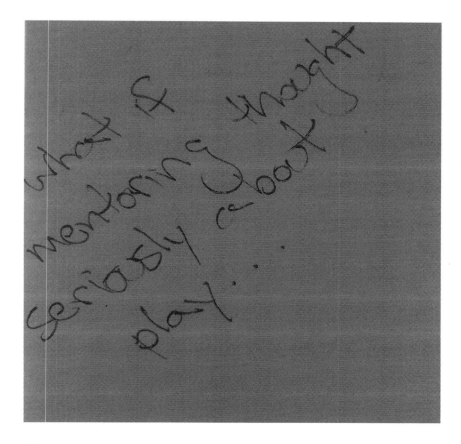

We suggest that speculative writers delve into words, rub words on their bodies, drink words with abandon, and slur their words whenever possible. Perhaps once we turn towards words as events themselves, we can enhance textual experience beyond representation. Can we engage in more play with words, and allow words to engage in more play with us? How might a page of play look, feel, smell, and taste? How might word-playing diverge from wordsmithing? How might word play allow exploration of temporal immanence? What words will we wear when we write?

Event ∞

Temporal uncertainty of experiencing events:

not being able to follow their mental variety and shifts.

Entering the space of drifting thoughts, visions and ideas,

wrapped in the possibility

of exploring and defying

the widest extension of the discussion.

Tenses create tensions:

thinking is bound by the temporality of language.

How would discussion-events expand if they were time-free?

Just as the temporality of *to be* has multiple possibilities:

what was, still is, in a different being, through be-coming, and will always be.

So does discussing.

Gerund can be an object, subject or complement of a verb.

Similar to a noun but implying dynamics and process.

Discussing has only relational time.

May-be.

Event ♪♫♩♫♯

How often do we get to encourage others to put something forth in the world that will inspire?

That life force that momentarily quickens inside of us, as Martha Graham (as cited in De Mille, 1991) says, the world will not have if we don't act upon it, bringing it into being?

When we come to a new institution, we are sometimes put through "training" that teaches us to view future students through an atmosphere of suspicion, as potential cheaters with ever-evolving online storehouses of plagiarized test banks and essays, or sloths trying to emerge with top grades with as little effort as possible.

Why are we not encouraged to consider them as creators in need of tools and a bit of guidance to dare to believe that they can bring those quickenings within them to light?

What does it do to a student body to caution its educators to police against the least desirable outcome rather than to keep an eye out for how to nurture flickers of the best possible outcome?

Event 9

The event demands the unorthodox and nonsensical. "Alice becomes larger . . . by the same token . . . she becomes smaller . . . this is the simultaneity of a becoming whose characteristic is to elude the present" (Deleuze, 1990, p. 1). The mentee-mentor becomes larger and, by the same token, smaller. The mentee-mentor (subjectivities) becomes paradox.

Contingency and chance are some of the event's defining features. The event doesn't simply spoil the preplanned and the systematized but creates their very conditions: "My wound existed before me; I was born to embody it" (Deleuze, 1994, p. 159). The mentor and mentee are planned always after the fact – after the wound: The job, the research trajectory, the outline of the paper, the methods, the procedures, the schedule, the course outline orders the world after the fact. To engage the event of mentoring means holding space for contingency and chance, the unorthodox and the nonsensical. Poetry as critique. Data as apparel.

What if our styles of apparel differ? Must I wear what is considered business casual data to survive?

Interview as slow-motion mentoring. Language as monstrous. Method as leaky mentor-mentees. Politics as mentoring carnival. Matter as immaterial. Observation as participation. Witnessing as wit(h)ness-ing. Text as mentoring texture. Pedagogy as implosion. Mentoring time as sugar cube. Root as rhizome. Writing as dirt. Research as mentoring creation. The events demand *amor fati*, eternal mentoring transformation.

Event 3.16.2020

Event contagion.
Infect minds and bodies, home and sleep.
Resist in vain. Partition, bind, discipline.
The qualitative – the quantitative; the sciences – the humanities; the experts – the amateurs.
Return uncertainty.
Enact surveillance, Enact inspection.
The exam, the review, the measurement, the responsibility, the timeline.
Return the uncounted.
Call for quarantine, call for exclusion
The library, the archive, the cubicle, the office
Return the contagion.
Enliven, inactivate, enrich.
Return life of the middle.
Eternal return the event.
"A will willing itself through all change, a power opposed to law, an interior of the earth opposed to the laws of its surface" (Deleuze, 1994, p. 6).
Call the event hope, not optimism.

Viral load is also key to event contagion.
The more bodies involved, the greater the load grows.
If we isolate ourselves, our loads likewise diminish,
and our habitual immune systems will strike the new from our systems
in favor of the safe.

When we are together, haecceities emerge
through an interior's leaning toward the exterior (Wyatt et al., 2011)
and we become implicated with one another.
There is no distinction between one and another,

> periphery and center (Halsey, 2008),
> belonging and nonbelonging.
> They thrive on speed and intensity.

Prepare for the worst but hope for the best.

. . . .

Within a mentoring relationship, the actual is so often confined to what we allow ourselves to voice directly, to a mentor-mentee, or to others outside of the mentor-mentee relationship. Immanent arisings also linger within the hinted, bodily encounters, through facial expressions and tone of voice that alludes to all that goes unvoiced. What if something (care, a realization, an idea, and affect) has already materialized within virtual mentor-mentee relations, manifested through difference, and adding to its appearances, present for one relation while absent for another?

. . . .

Event 202020–3

> *How to leap from philosophy, often written so impersonally, as a springboard*
> *for our intimate encounters with mentorship?*
> *Those lofty propositions of melding into one another, co-existing . . .*
> *When we met so that all of our ideas could breathe together, we felt an*
> *adrenaline rushing through us, pushing us from one point to another,*
> *it all felt so expansive, we didn't know where to begin.*
> *We knew that we did not want to euphemize my experiences of being*
> *mentor-mentored.*
> *We live in the real, not only in the aspiring,*
> *and we carry its reverberations with us every day*
> *It had to be philosophy as lived experience.*
> *Maybe a philosophy of living qualitative inquiry.*
> *To create any other form feels like a betrayal.*
> *Could we experiment and re-think qualitative inquiry practices and our living*
> *mentoring events? Where might we go together?*

What did you take with you from our last meeting?

Not sure we had an agenda and a take-away message. Maybe I was not
sure where these inquiries will end. How might I express that without
sounding endless-ly relativist?
We are nurturing ourselves with rationed slivers of those affirmations of
promise that you breathed to us last time . . .

I don't want to breathe to you. I resist. I resist.

Trying to set aside the assumptions you made about our work and the
scope of our vision, because, well, that's not really the point, is it? It was
a reference point that you brought with you into our conversation, that
hovered nearby, until several passes back and forth between us finally dis-
pelled it.

We wait for the moments when we can sway against one another . . . flow
into one another,
    not because we help you remember something
        or because you facilitate a new connection for us to reach out to,
but for us to arrive at a place of greater understanding . . . of insight . . . by
letting/
                /by letting understandings happen and un-happen.
                By letting understandings of inquiry processes,
                                  theories,
           role of subjectivities and ethics disappear
                as soon as they are created

      our eddies of
        experience and knowledge
        flow together.
              Immediately to disperse again.
   Like a seedling
      planted in the wrong climate
          in the wrong context,
          research design,
       within inaccurate methods,
    and insufficient data representation.
      Questionable ethics.
needing the proper light and temperature to continue to germinate.
    we sense you feeling us.

Trying but not achieving and never completing.

We try to learn where you are coming from ~
We read your autobiography,
　　　your methodologies,
　　　　　　your personal rants and petition shares on social media,
　　　　　　　　your engagements and performances at conferences . . .

But how could you since when asked I (subject of "I") never created them?

We go to your favorite yoga class together ~
You take us to a museum exhibit that feels connected with budding ideas
for our dissertation work ~
You gift us books that you think will resonate with us ~

Assumptions, misjudgments at best.

We bring you spices from our family members' homelands and a curtain to
block the glare of sunlight that floods your seat from your office window.

Sometimes, we escape from the reach of power that knowledge brings . . .
for a conversation . . . during a period of silent appreciation . . . before you
make a joke based on a reference that we do not catch,

in that moment,
　　　　　　　our presence
diminishes
　　　　　and we wait for the next moment
when we might color in its contours a little more visibly.

References

Al Makhamreh, M., & Stockley, D. (2019). Mentorship and well-being: Examining doc-
　　toral students' lived experiences in doctoral supervision context. *International Journal of
　　Mentoring and Coaching in Education, 9*(1), 1–20.
Arnesson, K., & Albinsson, G. (2017). Mentorship: A pedagogical method for integration of
　　theory and practice in higher education. *Nordic Journal of Studies in Educational Policy,
　　3*(3), 202–217.
Blackwell, J. E. (1989). Mentoring: An action strategy for increasing minority faculty. *Aca-
　　deme, 75*(5), 8–14.

Brondyk, S., & Searby, L. (2013). Best practices in mentoring: Complexities and possibilities. *International Journal of Mentoring and Coaching in Education, 2*(3), 189–203.

Chan, A. W. (2008). Mentoring ethnic minority, pre-doctoral students: An analysis of key mentor practices. *Mentoring & Tutoring: Partnerships in Learning, 16*(3), 263–277.

Council of Graduate Schools. (2010). *Ph.D. completion and attrition: Policies and practices to promote student success.* Council of Graduate Schools.

Dedrick, R. F., & Watson, F. (2002). Mentoring needs of female, minority, and international graduate students: A content analysis of academic research guides and related print material. *Mentoring & Tutoring: Partnerships in Learning, 10*(3), 275–289.

Deleuze, G. (1990). *The logic of sense* (M. Lester, Trans.). Columbia University Press.

Deleuze, G. (1994). *Difference and repetition* (P. Patton, Trans.). Columbia University Press.

Deleuze, G. (1995). *Negotiations 1972–1990* (Joughin, M., Trans.). Columbia University Press.

Deleuze, G., & Guattari, F. (1983). *Anti-Oedipus: Capitalism and schizophrenia* (R. Hurley, M. Seem, & H. R. Lane, Trans.). University of Minnesota Press.

Deleuze, G., & Guattari, F. (1994). *What is philosophy?* (H. Tomlinson & G. Burchell, Trans.). Columbia University Press.

De Mille, A. (1991). *Martha: The life and work of Martha Graham.* Vintage.

Felder, P. (2010). On doctoral student development: Exploring faculty mentoring in the shaping of African American doctoral student success. *The Qualitative Report, 15*(2), 455–474.

Grant, C. M., & Ghee, S. (2015). Mentoring 101: Advancing African-American women faculty and doctoral students in predominantly White institutions. *International Journal of Qualitative Studies in Education, 28*(7), 759–785.

Hall, L. A., & Burns, L. D. (2009). Identity development and mentoring in doctoral education. *Harvard Education Review, 79*(1), 49–70.

Halsey, M. (2008). Molar ecology: What can the (full) body of an eco-tourist do? In A. Hickey-Moody & P. Malins (Eds.), *Deleuzian encounters: Studies in contemporary social issues* (pp. 135–150). Palgrave Macmillan.

Jacobi, M. (1991). Mentoring and undergraduate academic success: A literature review. *Review of Educational Research, 61*(4), 505–532.

Kram, K. E. (1988). *Mentoring at work: Developmental relationships in organizational life.* University Press of America.

Lyons, W., Scroggins, D., & Rule, P. B. (1990). The mentor in graduate education. *Studies in Higher Education, 15*(3), 277–285.

Manning, E. (2013). *Always more than one: Individuation's dance.* Duke University Press.

Mansfield, K. C., Welton, A., Lee, P., & Young, M. D. (2010). The lived experiences of female educational leadership doctoral students. *Journal of Educational Administration, 48*(6), 727–740.

Mullen, C. A., Fish, V. L., & Hutinger, J. L. (2010). Mentoring doctoral students through scholastic engagement: Adult learning principles in action. *Journal of Further and Higher Education, 34*(2), 179–197.

Noy, S., & Ray, R. (2012). Graduate students' perceptions of their advisors: Is there a systematic disadvantage in mentorship? *The Journal of Higher Education, 83*(6), 876–914.

Roberts, L., Tinari, C. M., & Bandlow, R. (2019). An effective doctoral student mentor wears many hats and asks many questions. *International Journal of Doctoral Studies, 14*, 133–159.

Rose, G. L. (2005). Group differences in graduate students' concepts of the ideal mentor. *Research in Higher Education, 46*(1), 53–80.

Samier, E., & Fraser, S. (2000). Public administration mentorship: Conceptual and pragmatic considerations. *Journal of Educational Administration, 38*(81), 83–101.

Shaviro, S. (2014). *The universe of things: On speculative realism.* University of Minnesota Press.

Wang, T., & Li, L. Y. (2011). "Tell me what to do" vs. "guide me through it": Feedback experiences of international doctoral students. *Active Learning in Higher Education, 12*(2), 101–112.

Whitehead, A. N. (1967). *Adventure of ideas.* The Free Press.

Woloshyn, V., Savage, M. J., Ratkovic, S., Hands, C., & Martinovic, D. (2019). Exploring professors' experiences supporting graduate student well-being in Ontario faculties of education. *International Journal of Mentoring and Coaching in Education, 8*(4), 397–411.

Wyatt, J., Gale, K., Gannon, S., & Davies, B. (2011). *Deleuze & collaborative writing: An immanent plane of composition.* Peter Lang.

Yob, I., & Crawford, L. (2012). Conceptual framework for mentoring doctoral students. *Higher Learning Research Communications, 2*(2), 34–47.

Before the Beginning

A poem for my mentor, Dr. Susan Copeland

Sharon Head

Before the beginning, before the absent turning from a plain path.
Before the thought that begat the word,
Before the gathering of air in my lungs,
Before the push of the diaphragm that caused pooled breath to rise,
Before the atonal, awkward flow past tooth, tongue and lip,
Before pent-up, self-denied dreams shaping and misshaping the words themselves,
Before the beginning
Was the space held for me by one who could only be called:

      Mentor through her prevenient belief in me,
      Mentor in her careful scribbling of notes as I spoke,
      Mentor through her act of holding space for me

To speak
Before the beginning, before the word,
Before the question that forced the thought that birthed the word,
Before the inhalation and slight holding in my lower lungs,
Before the contraction that caused the gathered song to rise,
Before that central muscle pounded down fear and
Before words became sentences and occasionally cohered,
Before I was a student, a candidate, a researcher, a done-dissertator
She held the space and waited, before my beginning.

 DOI: 10.4324/9781003022558-27

INDEX